The Impact of Artificial Intelligence on Human Rights Legislation

John-Stewart Gordon

The Impact of Artificial Intelligence on Human Rights Legislation

A Plea for an AI Convention

palgrave
macmillan

John-Stewart Gordon
Lithuanian University of Health Sciences
Kaunas, Lithuania

ISBN 978-3-031-31390-5   ISBN 978-3-031-31388-2   (eBook)
https://doi.org/10.1007/978-3-031-31388-2

This Palgrave Macmillan imprint is published by the registered company Springer Nature Switzerland AG.
The registered company address is: Gewerbestrasse 11, 6330 Cham, Switzerland

To
My wife

CONTENTS

About the Author

John-Stewart Gordon is an adjunct professor at Lithuanian University of Health Sciences (LSMU) in Kaunas, Lithuania (since 2022). Before, he was Full Professor of Philosophy (2015–2022), chief researcher at the Faculty of Law (2018–2022), founder and head of the Research Cluster for Applied Ethics (2016–2022), and principal investigator of the EU-funded research project "Integration Study on Future Law, Ethics, and Smart Technologies" (2017–2022) at Vytautas Magnus University in Kaunas, Lithuania. Furthermore, he is an associate editor at *AI & Society* (since 2020) and a member of the editorial boards of *Bioethics* (since 2007), *AI & Society* (since 2018), and the *Baltic Journal of Law & Politics* (2018–2022), and he has been an area-editor and board member of the Internet Encyclopedia of Philosophy (2007–2014). Furthermore, he is the general editor of the book series *Philosophy and Human Rights*. John spent extended research stays at the universities of Queen's, Oxford (multiple times), Toronto, Maynooth, Tallinn (multiple times), the Karolinska Institutet in Stockholm, and the Academy of International Affairs NRW in Bonn. He has written and edited numerous books in the context of practical philosophy and published many peer-reviewed articles and special issues in leading international journals and encyclopaedias.

LIST OF TABLES

List of Tables

CHAPTER 1

General Introduction

Abstract In this chapter, the author delves into the complex topic of robot rights, outlining the book's central theme and primary argument. The discussion begins with an examination of two prominent sceptical viewpoints on robot rights, attributed to scholars A. Birhane, J. van Dijk, and L. Floridi. Birhane and van Dijk champion human exceptionalism, emphasising human uniqueness and superiority, thus casting doubt on the necessity for robot rights. In contrast, Floridi's mindless morality concept contests the idea of robots having moral agency, questioning the legitimacy of granting them rights. The author argues that these sceptical perspectives are ultimately unconvincing. To facilitate a more insightful and nuanced dialogue on robot rights, it is crucial to remain open-minded when

The book was written during my research stay at the Academy of International Affairs in NRW, Bonn, in 2022. I would like to express my sincere gratitude for the opportunity to carry out the project and the excellent on-site support provided. In addition, I express my gratitude to the attendees of my book content presentation for their insightful feedback. Lastly, I would like to thank Palgrave Macmillan for including the book in their portfolio.

J.-S. Gordon, *The Impact of Artificial Intelligence on Human Rights Legislation*, https://doi.org/10.1007/978-3-031-31388-2_1

contemplating the ethical considerations and potential ramifications of general artificial intelligence and superintelligence. By reorienting the conversation, the author invites readers to reassess preconceived beliefs about robots' roles, responsibilities, and the evolving dynamics between humans and advanced technologies.

Keywords General AI • Superintelligence • Human exceptionalism • Mindless morality view • Open-mindedness • Robot rights

The unmatched technological achievements in artificial intelligence (AI), robotics, computer science, and related fields over the last few decades can be considered a success story (Gordon & Nyholm, 2021; Müller, 2020). The technological sophistication has been so ground-breaking in various types of applications that many experts believe that we will see, at some point or another, the emergence of general AI (AGI) and, eventually, superintelligence (Bostrom, 2014; Kurzweil, 2005).

The implementation of smart technologies in our world is comprehensive. It already causes substantial moral and legal problems in various fields, such as law (Angwin et al., 2016) and the banking sector (Ludwig, 2015), where the issue of *machine bias* has arisen, with a detrimental impact on many people's lives. In recent years, experts have examined numerous issues concerning violations of fundamental human rights, such as the right to equality and non-discrimination in the context of AI and its application (Bacciarelli et al., 2018). Undoubtedly, AI offers excellent opportunities and helps humanity rise to a level never dreamed of, but it also presents extreme risks that must be minimised.

Technologies are never simply "neutral"; they can be used for positive benefit or immoral deeds. When invented, the sword was used not only to defend people's lives but also to conquer countries and enslave other humans illegally. Likewise, AI can promote human flourishing and well-being but can also violate human rights, such as the right to privacy, equality, and non-discrimination.

Furthermore, even if there is no intent to use AI for immoral deeds, the software (if programmed poorly) and data (if unprepared) can violate people's human rights and cause substantial injustice (Raso et al., 2018).

More and more experts are realising that these issues are of utmost significance and must be adequately addressed to avoid undesirable outcomes. Given the far-reaching consequences of AI for people's lives (e.g. machine bias in law, autonomous drones in warfare, AI-assisted policing, and China's social credit point system), some scholars have examined the impact of AI on human rights in greater detail (Gordon, 2022a; Gordon & Pasvenskienė, 2021; Livingston & Risse, 2019; Risse, 2019). Even though we have already reached a level where AI's negative effects on human rights are visible, the emergence of AGI and superintelligence could be even worse for humanity.

Scholars such as Nick Bostrom (2014) and Toby Ord (2020) have repeatedly warned about the likely emergence of superintelligent machines, which could cause the extinction of humanity if we fail to align the values and norms of these superintelligent machines with accepted human moral values and norms. And even if superintelligent machines are *benevolent*, other serious threats, such as conflicts over resources, existential boredom (Bloch, 1954), human "enfeeblement" (Russell, 2019), "de-skilling" in different areas of human life (Vallor, 2015; Vallor, 2016), and the future of human work (Danaher, 2019), could arise. These problems are significant, and we should think about solutions now and not when it is too late. Meanwhile, there is evidence that humans might violate so-called machine rights once such machines exist, especially in the context of sex (Gordon & Nyholm, 2022; Levy, 2008) and companion robots (Nyholm, 2020).

1.1 General Line of Argumentation

This book examines the impact of AI on human rights (by focusing on potential risks) and human rights legislation and proposes creating a *Universal Convention for the Rights of AI Systems* (AI

Convention). The second chapter (after this general introduction) introduces the concept of AI. It offers an initial overview concerning important elements of AI that are relevant to the book's general theme, including a working definition of AI, a section on the relation between AI and ethics, and a discussion of the concepts of AGI and superintelligence.

The third chapter provides a brief overview of the concept and the main features of human rights. It distinguishes between moral and legal human rights while highlighting the fact that international legal human rights practice is central to the agenda of creating an AI Convention. Admittedly, the exact relationship between moral and legal human rights remains unresolved. Nevertheless, the legal human rights discourse depends on moral human rights and cannot be reasonably substantiated without this moral foundation. Furthermore, I argue that the concept of personhood is central to the entitlement to human rights.

The fourth chapter examines the relationship between AI and human rights, especially AI's (negative) impact on legal human rights. Against this background, two zones of conflict (machines violate human rights and humans violate "machine rights") are presented, along with their impact on human rights legislation. They are associated with three different time frames (early twenty-first century, 2040–2100, and starting from the 2100s). This chapter identified current, mid-term, and long-term human rights violations in the context of AI. The section on how *humans violate "machine rights"* is particularly significant because experts rarely consider this topic and underestimate its importance for a peaceful relationship between humans and AGI (or superintelligent machines) once they exist.

The fifth chapter mainly concerns how one can best *protect* superintelligent machines from harm done by humans. The best way, I contend, is to create a *Universal Convention for the Rights of AI Systems* so as to provide these highly sophisticated and technologically advanced machines with adequate protection. The language of human rights has great rhetorical, moral, and legal force in addition to a long-established institutional place in international law. The sections of this chapter stress the significance of moral status and related rights. The chapter argues that AI machines will not only be entitled to fundamental rights but also have substantial moral and legal duties towards humans and humanity, which is decisive concerning fears of "robot revolutions". If superintelligent machines acknowledge and appreciate that they are protected by an AI Convention, one can hope that they will recognise and respect their duties towards humans and humanity (e.g. the Peter Parker principle[1]).

The final chapter presents some closing remarks.

1.2   The Caveat

Some experts might argue that my ideas utterly depend on the controversial view that AGI and superintelligence will emerge at some point (e.g. Searle, 1980). That is correct. The claim that we will see artificial machines that might match or surpass human capabilities is uncertain but not impossible. Others believe, however, that there is no *principal* reason why this will not eventually be the case (Chalmers, 2010); hence, we should be prepared for this scenario before it happens (Bostrom, 2014; Ord, 2020). In general, *the precautionary principle* recommends that we think things through before they exist and gauge the possibility of their existence. In this situation, that principle calls for a moral and legal assessment of the potential impact of AGI and superintelligence on human rights practice.

Undoubtedly, at this moment, many scholars are working on *AI regulation* and how to mitigate the effects of damage caused by AI (e.g. autonomous transportation and drones in warfare). My proposal is different and more audacious in arguing that we should have an AI Convention that protects AGI and superintelligent machines once they exist. This step is warranted not only because, given their moral status, they would be entitled to moral and legal rights (Gordon, 2020) but also because of the argument that

---

[1] The PPP states that "with great power comes great responsibility".

protecting them against harm gives them *a strong reason to accept moral and legal duties towards humans and humanity* as well. The following chapters will develop this line of reasoning. I hope they will stimulate a general discussion concerning the impact of AI on human rights legislation.

1.3   EXTREME SCEPTICISM AND ROBOT RIGHTS

The concept of "granting" rights to robots, which relies on future technological advancements, is frequently met with significant scepticism from scholars across various disciplines. Some, such as Birhane and van Dijk (2020), contend that advocating for robot rights based on speculative visions of sentient machines is little more than philosophical speculation ("armchair philosophy"). In their view, the debate over robot rights is a misguided concern limited primarily to more privileged societies. They argue that this focus on robot rights is rooted in a sense of technological arrogance or an overly optimistic view of the potential of the technology industry.

The perspective of extreme sceptics regarding robot rights can be frustrating for several reasons. First, scholars from disciplines outside philosophy and ethics sometimes make normative arguments that are misleading and logically flawed due to their shallow understanding of essential ethical theories, concepts, and debates. This approach violates academic standards and prevents an unbiased and honest discussion. For example, Birhane and van Dijk's paper (2020) exhibits this weakness.

Second, some philosophers and ethicists, such as Luciano Floridi, may also be extreme sceptics regarding robot rights. Floridi, in his online essay "Should We Be Afraid of AI?" (2016), argues that although true AI is logically possible, it is "utterly implausible". This viewpoint is flawed because it underestimates the potential for significant technological advancements in the future. The feasibility of true AI is ultimately an empirical question, and the lack of imagination about future developments is a concern.

In the following discussion, I will briefly summarise Birhane and van Dijk's main argument against robot rights and then comment on it. I will also touch on Floridi's view, but only briefly.

1.3.1 A Plea for Human Exceptionalism

Cognitive scientist Abeba Birhane and design researcher Jelle van Dijk argue that the debate surrounding robot rights is futile because robots are not and, for principled reasons (namely, that they are only the "mediators" of human beings), never will be entities that could be "granted" rights. Instead, they suggest focusing on the negative impact that smart technologies have on human beings, particularly those from disadvantaged groups who face issues such as machine bias and discrimination (2020, pp. 4–5). According to Birhane and van Dijk, prioritising the discussion of robot rights over addressing these negative impacts could result in inflicting harm on these groups (2020, p. 4).

Birhane and van Dijk's stance on robot rights differs from that of proponents such as David Gunkel (2012, 2018) and Mark Coeckelbergh (2014), who argue for taking the concept of robot rights seriously. Despite starting from the same normative viewpoint, which is grounded in a post-Cartesian social-relational approach and a phenomenologically inspired worldview based on the social nature of human beings, Birhane and van Dijk (2020, p. 2) draw a different conclusion when it comes to the inclusion of machines.

Gunkel and Coeckelbergh believe that moral status is not exclusive to human beings but can also be extended to other entities, such as machines, through social relations. In contrast, Birhane and van Dijk are sceptical that the social-relational approach can be extended to include machines. They argue that human beings and their welfare should be the primary concern (2020, p. 1, 3) and that robots, being machines and not human beings, are not the kind of entities that deserve rights (2020, p. 2). They criticise the focus on robot rights and advocate for dealing with the negative impact of AI on human welfare, particularly of oppressed groups. For example, whereas Gunkel's starting point is based on the idea of "otherness" and the social nature of entities beyond humans, Birhane and van Dijk prioritise human beings and their welfare (a human-centred approach). This fundamental difference in perspective undergirds their extreme scepticism towards robot rights.

There are two reasons why the idea that only *humans* have rights by virtue of their *humanity* is problematic according to the standard view in ethics and analytic contemporary moral philosophy. First, this view is flawed according to Hume's Law,[2] which states that no normative conclusions (e.g. whether an entity deserves moral rights) can be drawn from empirical facts (e.g. whether the entity possesses human DNA) alone.

Second, the recognition of rights for humans is based not simply on their biology but on the concept of personhood. To argue otherwise is to engage in speciesism, a form of discrimination similar to sexism and racism (Singer, 2009). By privileging humans over other beings, such as robots, Birhane and van Dijk's argument is misleading and unjustified based on a human-comes-first doctrine. The concept of personhood, not biology, should be used as the basis for recognising rights. This objection has been presented elsewhere to demonstrate why human exceptionalism is unfounded (Gordon, 2022a; Gordon, 2022b; Gordon & Pasvenskienė, 2021; Kamm, 2007; Singer, 2009).

Third, in addition to the above reasons for recognising the importance of personhood over mere humanity in determining moral status and rights, it is crucial to establish species-independent criteria for personhood so as to avoid the issue of speciesism. Recently, Kestutis Mosakas (2021) has argued for the consciousness criterion as the primary determinant of moral status and rights for all beings, regardless of whether they are human or non-human.

---

[2] In essence, Hume's Law argues that one cannot derive an "ought" statement from an "is" statement. In other words, it is impossible to move from a statement of fact or description of what is the case to a prescriptive statement about what ought to be the case.

This provides a compelling counterargument to the social-relational approach in machine ethics.

Fourth, Birhane and van Dijk have an erroneous understanding of moral rights. They believe that rights are "granted" and that robots will never deserve such rights in the first place. However, this view is flawed because the strength of moral rights does not depend on anyone "granting" them. If moral rights were dependent on recognition, this could further marginalise groups who have previously been oppressed. One intrinsic aspect of moral rights is that all entities possess them if they meet objective criteria, irrespective of whether other human beings recognise them (Francione, 2009; Gordon, 2021; Gordon & Pasvenskiené, 2021; Singer, 2009).

Fifth, the claim that a debate on robot rights contributes to harming individuals and groups is misguided. Such harm would result only if all research were solely focused on robot rights, which is not the case. Working on the issue of robot rights does not detract from examining other ethical issues surrounding AI. It is not appropriate to criticise colleagues interested in robot rights and dismiss their academic interests when they believe that this is a significant issue that will have moral implications. Even if debating the issue of robot rights does not pay off in any applications of concrete, practical value, it is unfair to belittle researchers for exploring it. This does not mean that the negative impact of AI on humans should be ignored. In fact, examining the negative impact of AI on humans, especially in the area of machine bias and discrimination, should be pursued along with debating robot rights. The two issues are not mutually exclusive, that is, there is no zero-sum situation.

1.3.2   The Mindless Morality View

Luciano Floridi, a philosopher and expert in AI, has a different perspective and does not believe that discussing "robot rights" is necessary. He claims the following:

> True AI is not logically impossible, but it is utterly implausible. We have no idea how we might begin to engineer it, not least because

we have very little understanding of how our own brains and intelligence work. This means that we should not lose sleep over the possible appearance of some ultraintelligence. What really matters is that the increasing presence of ever-smarter technologies is having huge effects on how we conceive of ourselves, the world, and our interactions. The point is not that our machines are conscious, or intelligent, or able to know something as we do. They are not. (Floridi, 2016)

While it is true that current AI systems do not possess consciousness or human-like intelligence, it is essential to consider the potential emergence of superintelligent AI and its possible implications. Even though machines have yet to achieve superintelligence, one cannot simply ignore the possibility; rather, we must be prepared to address the ethical, legal, and socio-political implications if and when it does arise. Therefore, it is essential to contemplate the potential consequences of superintelligence and to provide normative guidance for the possibility of machines, such as robots, being *entitled* to moral and legal rights.

Floridi's reluctance to discuss "robot rights" is based on his so-called mindless morality view (Floridi, 1999; Floridi & Sanders, 2004), which embodies a different ethical approach from others. However, given the recent rapid technological developments, it may not be prudent to dismiss the possibility of achieving "true AI" as "utterly implausible". Instead, common sense suggests that we should prepare for the potential consequences of superintelligent AI and consider how to offer normative guidance if humans are confronted by highly advanced AI machines of utmost sophistication.

Floridi's mindless morality view emphasises the role of automated decision-making systems in shaping ethical behaviour. He argues that advances in technology are making it increasingly possible to automate decision-making processes, including those related to ethical and moral questions. For example, self-driving cars may need to decide whom to prioritise in a dangerous situation, such as choosing between hitting a pedestrian or swerving off the road and potentially harming the car's occupants. According to Floridi, in cases such as these, it may be possible to develop algorithms that can make ethically justifiable decisions. These algorithms would be based on predefined ethical principles and would operate independently of human input or oversight. He claims that such automated systems can be seen as a form of "mindless morality" because they do not require conscious moral deliberation or decision-making.

In other words, one does not necessarily need to aim for "machine consciousness" to develop advanced AI systems. This approach, which Floridi espouses, is parsimonious because it avoids the need for complex assumptions such as self-awareness and multitasking that are relevant for proponents of robot rights (see Sect. 2.4). However, as AI technology continues to advance at an unprecedented pace, the potential for a superintelligence that surpasses human intelligence becomes increasingly plausible. Therefore, it is crucial to consider the ethical, legal, and socio-political implications of robot rights sooner rather than later.

1.3.3 Open-mindedness

With rapid technological advancements, the potential for artificial intelligence to exceed human intelligence and create superintelligence is becoming increasingly feasible. As AI systems continue to evolve and become more sophisticated, their ability to learn and adapt may lead to the development of a superintelligence that surpasses human comprehension. The intelligence explosion (Good, 1965), which occurs as AI systems improve themselves, is a crucial factor in the emergence of superintelligence (see Sect. 2.5). Theoretical foundations, such as recursive self-improvement,[3] support the idea that superintelligence is possible and could be achieved through technological progress. In general, these concepts suggest

---

[3] Recursive self-improvement implies that an AI system with sufficient intelligence could enhance its own source code, thereby significantly increasing its overall intelligence.

that superintelligence could be a realistic possibility in the future, although the precise time frame remains uncertain. The potential for the creation of superintelligence through advanced AI technologies underscores the importance of carefully considering such developments' ethical, legal, and socio-political implications.

Instead of being overly sceptical about the possibility of superintelligent robots emerging, I believe it is essential to approach the debate with an open-minded perspective. The possible future existence of superintelligent robots is an empirical question that cannot be prematurely dismissed in advance. Therefore, I advocate for a thoughtful and objective discussion that explores the possibility of their emergence with a critical yet curious mindset.

REFERENCES

Angwin, J., Larson, J., Mattu, S., & Kirchner, L. (2016). Machine Bias. *ProPublica*, May 23. https://www.propublica.org/article/machine-bias-risk-assessments-in-criminal-sentencing

Bacciarelli, A., Westby, J., Massé, E., Mitnick, D., Hidvegi, F., Adegoke, B., Kaltheuner, F., Jayaram, M., Córdova, Y., Barocas, S., & Isaac, W. (2018). The Toronto Declaration: Protecting the Right to Equality and Non-discrimination in Machine Learning Systems. *Amnesty International and Access Now*, 1–17.

Birhane, A., & van Dijk, J. (2020). Robot Rights? Let's Talk About Human Welfare Instead. AAAI/ACM Conference on AI, Ethics, and Society (AIES'20), February 7–8, 2020, New York, NY, USA, 1–7. https://doi.org/10.1145/3375627.3375855

Bloch, E. (1954). *Das Prinzip Hoffnung* (3 vols.). Suhrkamp.

Bostrom, N. (2014). *Superintelligence: Paths, Dangers, Strategies.* Oxford University Press.

Chalmers, D. (2010). The Singularity: A Philosophical Analysis. *Journal of Consciousness Studies, 17,* 7–65.

Coeckelbergh, M. (2014). The Moral Standing of Machines: Towards a Relational and Non-Cartesian Moral Hermeneutics. *Philosophy & Technology., 27*(1), 61–77.

Danaher, J. (2019). *Automation and Utopia.* Harvard University Press.

Floridi, L. (1999). Information Ethics: On the Philosophical Foundation of Computer Ethics. *Ethics and Information Technology, 1*(1), 37–56.

Floridi, L. (2016). Should We Be Afraid of AI? *Aeon* (Online Magazine). https://aeon.co/essays/true-ai-is-both-logically-possible-and-utterly-implausible

Floridi, L., & Sanders, J. W. (2004). On the Morality of Artificial Agents. *Minds and Machines, 14*(3), 349–379.

Francione, G. L. (2009). *Animals as Persons: Essays on the Abolition of Animal Exploitation.* Columbia University Press.

Good, I. J. (1965). Speculations Concerning the First Ultraintelligent Machine. *Advances in Computers, 6,* 31–88.

Gordon, J.-S. (2020). What Do We Owe to Intelligent Robots? *AI & Society, 35,* 209–223.

Gordon, J.-S. (2021). Artificial Moral and Legal Personhood. *AI & Society, 36*(2), 457–471.

Gordon, J.-S. (2022a). Are Superintelligent Robots Entitled to Human Rights? *Ratio, 35*(3), 181–193.

Gordon, J.-S. (2022b). The African Relational Account of Social Robots: A Step Back?. *Philosophy & Technology, 35* (Online First). https://doi.org/10.1007/s13347-022-00532-4

Gordon, J.-S., & Nyholm, S. (2021). The Ethics of Artificial Intelligence. *Internet Encyclopedia of Philosophy.*

Gordon, J.-S., & Nyholm, S. (2022). Kantianism and the Problem of Child Sex Robots. *Journal of Applied Philosophy, 39*(1), 132–147.

Gordon, J.-S., & Pasvenskienė, A. (2021). Human Rights for Robots? A Literature Review. *AI and Ethics, 1,* 579–591.

Gunkel, D. (2012). *The Machine Question. Critical Perspectives on AI, Robots, and Ethics.* MIT Press.

Gunkel, D. (2018). *Robot Rights.* MIT Press.

Kamm, F. (2007). *Intricate Ethics: Rights, Responsibilities, and Permissible Harm.* Oxford University Press.

Kurzweil, R. (2005). *The Singularity Is Near.* Penguin Books.

Levy, D. (2008). *Love and Sex with Robots.* Harper Perennial.

Livingston, S., & Risse, M. (2019). The Future Impact of Artificial Intelligence on Humans and Human Rights. *Ethics & International Affairs, 33*(2), 141–158.

Ludwig, S. (2015). Credit Scores in America Perpetuate Racial Injustice: Here's How. *The Guardian,* October 13. https://www.theguardian.com/commentisfree/2015/oct/13/your-credit-score-is-racist-heres-why

Mosakas, K. (2021). On the Moral Status of Social Robots: Considering the Consciousness Criterion. *AI & Society, 36*(2), 429–443.

Müller, V. C. (2020). Ethics of Artificial Intelligence and Robotics. *Stanford Encyclopedia of Philosophy*, https://plato.stanford.edu/entries/ethics-ai/

Nyholm, S. (2020). *Humans and Robots: Ethics, Agency, and Anthropomorphism*. Rowman and Littlefield.

Ord, T. (2020). *The Precipice: Existential Risk and the Future of Humanity*. Hachette Books.

Raso, F., Hilligoss, H., Krishnamurthy, V., Bavitz, C., & Levin, K. (2018). Artificial Intelligence and Human Rights: Opportunities and Risks. Berkman Klein Center for Internet and Society Research Publication. http://nrs.harvard.edu/urn-3:HUL.InstRepos:38021439

Risse, M. (2019). Human Rights and Artificial Intelligence: An Urgently Needed Agenda. *Human Rights Quarterly, 41*, 1–16.

Russell, S. (2019). *Human Compatible*. Viking Press.

Searle, J. R. (1980). Minds, Brains, and Programs. *Behavioural and Brain Sciences, 3*(3), 417–457.

Singer, P. (2009). Speciesism and Moral Status. *Metaphilosophy, 40*(3–4), 567–581.

Vallor, S. (2015). Moral Deskilling and Upskilling in a New Machine Age: Reflections on the Ambiguous Future of Character. *Philosophy & Technology, 28*(1), 107–124.

Vallor, S. (2016). *Technology and the Virtues: A Philosophical Guide to a Future Worth Wanting*. Oxford University Press.

CHAPTER 2

What Is Artificial Intelligence?

Abstract In this chapter, the author presents a concise introduction to artificial intelligence (AI) as a basis for subsequent discussions. A working definition of AI is provided, along with elucidation of the distinctions between weak and strong AI and an exploration of their potential ethical and legal ramifications. Furthermore, the concept of superintelligence is introduced and differentiated from strong AI. As AI grows more powerful, its influence on ethical, socio-political, and legal aspects becomes increasingly significant. Consequently, assessing AI's impact on human rights legislation is crucial to address potential violations involving both humans and AI machines.

Keywords Artificial intelligence • Ethics of AI • Weak and strong AI • Superintelligence

2.1 Introduction

This chapter provides a brief overview of the concept of artificial intelligence (AI), to lay the foundation for the following chapters. After offering a working definition of AI in the next Sect. 2.2, I will

J.-S. Gordon, *The Impact of Artificial Intelligence on Human Rights Legislation*, https://doi.org/10.1007/978-3-031-31388-2_2

briefly depict the differences between weak and strong AI and will draw some essential conclusions concerning their possible ethical and legal impact (Sect. 2.3). In Sect. 2.4, the concept of superintelligence will be introduced and distinguished from the notion of strong AI in more detail. The stronger the AI becomes, the more impact it has on ethical, socio-political, and legal issues, due to the increasing capabilities of AI machines and their general impact on the lives of human beings. Therefore, it is of utmost importance to determine the impact of AI on human rights legislation in more general terms due to the possibility of human rights violations, either because AI machines could violate the rights of human beings or because human beings might violate robot rights.

2.2   Defining Artificial Intelligence

The philosopher and director of the Turing Archive for the History of Computing at the University of Canterbury, B. J. Copeland (2020), describes AI as

> the ability of a digital computer or computer-controlled robot to perform tasks commonly associated with intelligent beings. The term is frequently applied to the project of developing systems endowed with the intellectual processes characteristic of humans, such as the ability to reason, discover meaning, generalize, or learn from past experience.

This working definition is broad enough to capture various ideas of what an intelligent system should be capable of, but also specific enough to pinpoint some of the essential features of the concept of intelligence. It is impossible to review all existing definitions of the concept of intelligence (and hence AI) here, given a large number of different proposals in the literature. There is no common understanding of the meaning of intelligence. However, the aforementioned definition provides a good starting point for our analysis.

Some scholars, such as the American philosopher John Searle (1980), argue that it is impossible to develop an artificial system comparable to human intelligence because the AI machine will never really "understand" what it is doing. His so-called Chinese room argument[1] was designed to question the possibility of strong (or general) AI needed for a machine *to think* like a human being. Other philosophers, such as David Chalmers (1996, chapter 9) believe, based on theories such as functionalism[2] and computationalism,[3] that it is possible, in principle, for a machine to become *conscious* enough to attain intelligence. They reject the claim that "intelligence" presupposes a particular substratum such as carbon-based beings (i.e. human beings). Rather, they believe that intelligence could evolve in silicon-based environments (i.e. machines). The essential issue is that the system must be sufficiently complex.

The tremendous technological developments in the context of AI, computer science, and informatics in recent times suggest that it may be only a matter of time before AI systems will become developed to the point that they will think like human beings, especially given the advances in machine learning (e.g. self-learning machines with deep neural network architecture and quantum processing). It seems reasonable to remain open-minded. The idea that there is a principal reason why strong AI (and superintelligence) will never exist seems somewhat far-fetched. Instead, whether we will see the advent of superintelligent machines is an *empirical* question.

---

[1] Searle imagines a person who sits in a closed room and receives letters in the Chinese language, containing questions. The person does not understand Chinese, but she can respond to the questions by using a guidebook instructing her to reply to questions (or Chinese letters) with this or that answer. The answers are correct, but the person in the room does not understand or know what she is asked or responding to. This situation, according to Searle, is similar to an AI system. It will never really "understand" the meaning of the questions and responses.

[2] Functionalism is the view that mental states are defined solely by their functional role, which can be performed via biological operations of the brain or artificial computations by a software.

[3] Computationalism is the view that cognition is the computation of functions, independent of whether the computation is biological or artificial.

2.3   Artificial Intelligence and Ethics

2.3.1   Three Developmental Periods

The idea that ethics comes into play only when strong AI systems exist is misleading. It is reasonable to distinguish between three periods concerning the development of AI and its impact on human society, especially when it comes to the ethical dimension (Gordon & Nyholm, 2021):

- Short-term (early twenty-first century)
- Mid-term (from the 2040s to the end of this century)
- Long-term (starting with the 2100s)

The first period shows that we already face severe problems—even without having attained strong AI—related to the use of AI systems. Some of the most fundamental problems involve machine bias in law. For example, many African Americans were denied parole in the US when a software called COMPAS[4] was used to assist the legal decision-making of judges, especially concerning the likelihood of prisoners becoming recidivists (Angwin et al., 2016). The AI systems used by some banks to determine the creditworthiness of loan applicants showed a racial bias (Ludwig, 2015). Furthermore, gender bias in hiring has occurred when an AI system was used (Dastin, 2018). Finally, the problem of responsibility in life-and-death decision arises regularly in the field of autonomous transportation, particularly concerning self-driving cars (Nyholm, 2018a, 2018b). These are only a few examples of ethical issues regarding weak AI.

The second time period—still not dealing with strong AI—presents numerous ethical issues, such as the growing concerns related to AI warfare, in which autonomous drones and so-called killer robots are playing critical roles (Asaro, 2012). Furthermore, as they attain greater capabilities, AI systems might cause mass

---

[4]This acronym stands for Correctional Offender Management Profiling for Alternative Sanctions.

unemployment with unforeseen consequences. In addition, widely used AI systems could undermine moral agency (de-skilling) and human autonomy (Danaher, 2019), especially in the context of AI governance when the organisation and management of public decision-making are done with the help of AI. This situation is further aggravated when we consider the so-called black box problem and the right to know the reasons for decisions. In such cases, AI use violates the principles of explainability and transparency in the decision-making process.

The third period is characterised by a high likelihood that strong AI and superintelligence will exist and will substantially influence human existence. Nick Bostrom (2014) and Toby Ord (2020) argue that there is an existential risk for humanity if superintelligence (i.e. machines that are more intelligent than human beings) becomes a reality. Some scholars believe that humanity might even face a machine revolution and that human beings could become either enslaved or extinct. Therefore, according to Bostrom, one needs to *align* the values and norms held by superintelligent machines with human values and norms.

Another line of reasoning concerns how humans should act towards machines with strong AI or superintelligent robots. Do such machines have a moral and legal status comparable to that of human beings? Are they entitled to the same or similar moral and legal rights independent of what humans think about this issue? Are these highly sophisticated machines non-human persons and hence entitled to fundamental rights? These complex issues must be addressed before we are confronted by the emergence of strong AI and superintelligence (Gordon, 2022).

The previous considerations show that there is a strong connection between AI and ethics, even before the arrival of strong AI and superintelligence. Furthermore, the cases mentioned above—even those in the short-term period—require legal action or at least an awareness that, for example, machine bias exists and could harm innocent people. The problem of machine bias, which often arises in racial and gender-related contexts but also affects people with impairments, is one of the most severe issues we currently face since it perpetuates and aggravates the harm done to people who are already disadvantaged socially. Machine bias is caused by defective algorithms, the use of unprepared (historical) data or outcome bias (Springer & Garcia-Gathright, 2018: 451). That means we must change our existing laws to ban the use of faulty software in sensitive areas, such as parole decisions, unless we can ensure that human beings are not harmed.

2.3.2   *The Ethics of Artificial Intelligence*

The significant ethical challenges that AI poses for human beings and their societies are well documented in the excellent introductions by John-Stewart Gordon and Sven Nyholm (2021), Vincent Müller (2020), Mark Coeckelbergh (2020), Janina Loh (2019), Catrin Misselhorn (2018), and David Gunkel (2012). The ethics of AI contains at least two different fields of application. The first field concerns ethical theories that attempt to solve any issues or problems between human beings and AI systems in general terms; the second field considers how machines should act. The latter area has been dubbed "machine ethics". One of the leading pioneers in this field, Susan Anderson (2011: 22), defines the goal of machine ethics as follows:

> to create a machine that follows an ideal ethical principle or set of principles in guiding its behaviour; in other words, it is guided by this principle, or these principles, in the decisions it makes about possible courses of action it could take. We can say, more simply, that this involves "adding an ethical dimension" to the machine.

In the following discussion, I will provide one example from each field to explain more appropriately what the distinction presented above means. On the theoretical level, one pioneer of the relational approach, David Gunkel (2012, 2018), argues that robots have moral status based on their social relations with human beings instead of applying objective criteria such as sapience (ability to reason) or sentience (ability to feel pain). According to Gunkel, the personal experience with the *Other* (i.e. the robot) is of utmost significance for his phenomenological approach. The general line of reasoning concerning the moral importance of the relational concept of personhood can be fleshed out as follows (see Gordon & Nyholm, 2021):

1. A social model of autonomy, under which autonomy is not defined individually but stands in the context of social relations.
2. Personhood is absolute and inherent in every entity as a social being (it does not come in degrees).
3. An interactionist model of personhood, according to which personhood is relational by nature (but not necessarily reciprocal) and defined in non-cognitivist terms.

Gordon and Nyholm (2021) comment, "The relational approach does not require the robot to be rational, intelligent or autonomous as an individual entity; instead, the social encounter with the robot is morally decisive. The moral standing of the robot is based on exactly this social encounter" (2.f.iii). This leads to the question of how one should treat AI machines against the background of Gunkel's proposal. Do AI machines or robots have any moral and legal rights? Are we allowed to use them however we want, or are there any moral and legal limits that human beings must observe when dealing with machines (Gordon & Nyholm, 2022)? These crucial questions must be answered.

The second example concerns the field of machine ethics. The science fiction writer Isaac Asimov was the first to introduce a code of ethics for machines, with four laws guiding their actions (Asimov's *Runaround* 1942, *Robots and Empire* 1986). The four laws are as follows:

0. A robot may not harm humanity or, by inaction, allow humanity to suffer harm.
1. A robot may not injure a human being or, through inaction, allow a human being to be harmed.

2. A robot must obey the orders given to it by human beings except where such orders would conflict with the first law.
3. A robot must protect its own existence as long as such protection does not conflict with the first or second law.

Many experts have considered whether Asimov's proposed code of ethics can be used for AI systems. Many of Asimov's stories portray the difficulties of the four laws when applied by robots, which should not be a surprise given that there were no exciting stories to write about in the first place (because the four laws are not sufficient).

However, the ethics of AI is—among other things—also concerned with the vital question of whether AI machines (from weak AI to strong AI to superintelligent machines) have moral and legal status and hence are entitled to moral and legal rights (Bostrom & Yudkowsky, 2014; Gordon, 2020). Usually, this question entails whether AI machines have moral and legal personhood based on morally relevant criteria such as autonomy, rationality, self-awareness, intelligence, or sentience (Gordon, 2021).

2.4   WEAK AND STRONG ARTIFICIAL INTELLIGENCE

It is common to distinguish between weak or narrow AI (NAI) and strong or general AI (AGI) when describing the capabilities of an AI system. The main difference between the two levels of AI systems is that NAI can deal with only one main task, such as playing chess or the complex Chinese game of Go (other examples include facial recognition systems used in policing or systems for autonomous transportation). For example, in earlier stages of computer development, chess programs such as IBM's DeepBlue used brute-force searches when competing against human players, even when DeepBlue defeated world champions such as Garry Kasparov in 1996 and 1997.

Nowadays, AI programs such as Google's DeepMind AlphaGo Zero operate quite differently, using self-play reinforcement learning based on deep neural network architecture. AlphaGo Zero was able to learn to play Go within just a few hours on its own (blank-slate learning) after being given only the basic rules of how to play Go and no further instructions. AlphaGo Zero eventually beat (by 100 to 0) a previous Go program, which was trained by human beings and defeated world champion Lee Sedol in 2017. This is a remarkable technological development.

In contrast, AGI does not exist now, and some experts, such as Searle (1980), believe we will never witness such a system. However, according to a recent poll, most experts in the relevant fields claim that we will eventually see the advent of AGI and superintelligent machines in the future (Müller & Bostrom, 2016). The step from an AI system that deals with only one task to an AI system capable of dealing with numerous different tasks is substantial and may take some time. One example of AGI could be an intelligent human-like robot (IHR), which is not necessarily superintelligent, but which deals with multiple jobs instead of just one task. But what are the requirements of such an AI system? I would suggest that this IHR should have at least seven different features to meet the standards for AGI:

1. Rationality: Ability to think.
2. Intelligence: Ability to reason, discover meaning, generalise, and learn from past experience.
3. Autonomy: Ability to act and decide utterly independent of human supervision.
4. Multitasking: Ability to do several different tasks.
5. Interactive: Ability to interact comprehensively with its environment and with human beings.
6. Self-awareness: AGI should be (at least minimally) self-aware.
7. Ethical: Moral reasoning and decision-making.

This list of necessary features provides a solid basis for an IHR with capabilities which could be used for different purposes. Many scholars believe such a robot should be considered a non-human person with associated moral and legal rights. This line of reasoning is supported by the two principles mentioned by Bostrom and Yudkowsky (2014):

> *The principle of substrate non-discrimination:* If two beings have the same functionality and the same conscious experience and differ only in the substrate of their implementation, then they have the same moral status.

> *The principle of ontogeny non-discrimination:* If two beings have the same functionality and the same conscious experience and differ only in how they came into existence, then they have the same moral status.

Indeed, it does not matter morally whether an entity is *carbon-based* or *silicon-based* if they have the same "functionality" and "conscious experience"; that is, it does not matter whether they were artificially built or born as humans. Both principles are essential steps towards a more inclusive approach which encompasses human beings and IHRs.

2.5   SUPERINTELLIGENCE

Superintelligence means that an AI system—for example, a human-like intelligent robot—is *smarter* than a human being. An AGI system differs from superintelligence by possibly reaching human intelligence at some point but would not necessarily exceed human cognitive capabilities.

Nick Bostrom (1998) defines superintelligence as "an intellect that is much smarter than the best human brains in practically every field, including scientific creativity, general wisdom and social skills". Whether such a superintelligent entity is also "conscious" and has "subjective experiences" remains an open question, according to Bostrom (1998). However, given its outstanding capabilities, it seems hard to imagine that such an entity would not be conscious or would not have subjective experiences. Ultimately, this is an empirical question that could not be answered in advance.

The idea of superintelligence was undoubtedly influenced by the notion of "intelligence explosion" introduced by statistician I. J. Good (1965) when he famously claimed:

Let an ultraintelligent machine be defined as a machine that can far surpass all the intellectual activities of any man however clever. Since the design of machines is one of these intellectual activities, an ultra-intelligent machine could design even better machines; there would then unquestionably be an "intelligence explosion", and the intelligence of man would be left far behind. Thus the first ultraintelligent machine is the last invention that man need ever make.

Technological developments since the 1960s have been breathtaking, and many experts believe that we will see the so-called technological singularity, that is, the point at which intelligent machines surpass human capabilities (Chalmers, 2010), in the next few decades (Vinge, 1993). The well-known futurist Ray Kurzweil predicts that we might encounter the "technological singularity" by 2045 (Kurzweil, 2005). Although I think we will see superintelligent machines in the future (superintelligence presupposes singularity), I do not believe it is wise to predict specific dates when such machines might emerge. However, it seems fair to say that we might encounter them within the next 100–150 years.

Bostrom (2014) and Ord (2020) are rightly concerned about how superintelligent machines will act towards human beings, especially if they have been abused or instrumentalised for human gain. The human–robot relationship[5] will become the cornerstone of how such machines view human beings and how they act. It seems evident that if one mistreats a person (including a non-human person), one should be cautious about how the person will develop and respond to the abuse later on—especially if the person is much[6] more intelligent and has more advanced capabilities than her "torturer". In other words, humans would dig their graves if they abused such superintelligent machines. The revanche could lead to

[5] For an interesting study on the relationship between humans and robots, see Nyholm (2020).

[6] Here, we must realise that we could be talking about a substantial difference, one that might resemble the difference between human beings and ants (or perhaps even more stark) and not a difference between human beings and, say, great apes.

a robot revolution, eventually destroying human life as we know it (Bostrom, 2014; Ord, 2020; Vinge, 1993).

One important solution is to ensure that superintelligent machines share our moral values and norms. That would be a good starting point for a fruitful relationship. Otherwise, humanity would run the risk of its doom.

2.6   Conclusions

This chapter has offered a brief overview of the concept of AI. Although it is not a comprehensive analysis, it should prepare the reader for the following chapters. The main takeaway message is that AI already impacts our moral and legal considerations and that these issues will become even more significant in the future. Accordingly, current human rights legislation should be revised to become more inclusive concerning intelligent systems once they emerge. But even features of today's AI systems (e.g. autonomous transportation and machine bias) already call for adjustments in our legislation. How to do this will be examined in the remaining chapters.

References

Anderson, S. L. (2011). Machine Metaethics. In M. Anderson & S. L. Anderson (Eds.), *Machine Ethics* (pp. 21–27). Cambridge University Press.

Angwin, J., Larson, J., Mattu, S., & Kirchner, L. (2016). Machine Bias. *ProPublica*, May 23. https://www.propublica.org/article/machine-bias-risk-assessments-in-criminal-sentencing

Asaro, P. (2012). On Banning Autonomous Weapon Systems: Human Rights, Automation, and the Dehumanization of Lethal Decision-Making. *International Review of the Red Cross, 94*, 687–709. https://doi.org/10.1017/S1816383112000768

Asimov, I. (1942). *Runaround: A Short Story*. Street and Smith.

Asimov, I. (1986). *Robots and Empire: The Classic Robot Novel*. HarperCollins.

Bostrom, N. (1998). How Long Before Superintelligence? *International Journal of Future Studies, 2*.

Bostrom, N. (2014). *Superintelligence: Paths, Dangers, Strategies*. Oxford University Press.

Bostrom, N., & Yudkowsky, E. (2014). The Ethics of Artificial Intelligence. In K. Frankish & W. M. Ramsey (Eds.), *The Cambridge Handbook of Artificial Intelligence* (pp. 316–334). Cambridge University Press.

Chalmers, D. (1996). *The Conscious Mind: In Search of a Fundamental Theory*. Oxford University Press.

Chalmers, D. (2010). The Singularity: A Philosophical Analysis. *Journal of Consciousness Studies, 17*, 7–65.

Coeckelbergh, M. (2020). *AI Ethics*. MIT Press.

Copeland, B. J. (2020). Artificial Intelligence. *Britannica.com*. https://www.britannica.com/technology/artificial-intelligence

Danaher, J. (2019). *Automation and Utopia*. Harvard University Press.

Dastin, J. (2018). Amazon Scraps Secret AI Recruiting Tool That Showed Bias Against Women. *Reuters*, October 10. https://www.reuters.com/article/us-amazon-com-jobs-automation-insight/amazon-scraps-secret-ai-recruiting-tool-that-showed-bias-against-women-id USKCN1MK08G

Good, I. J. (1965). Speculations Concerning the First Ultraintelligent Machine. *Advances in Computers, 6*, 31–88.

Gordon, J.-S. (2020). What Do We Owe to Intelligent Robots? *AI & Society, 35*, 209–223.

Gordon, J.-S. (2021). Artificial Moral and Legal Personhood. *AI & Society, 36*(2), 457–471.

Gordon, J.-S. (2022). Are Superintelligent Robots Entitled to Human Rights? *Ratio, 35*(3), 181–193.

Gordon, J.-S., & Nyholm, S. (2021). The Ethics of Artificial Intelligence. *Internet Encyclopedia of Philosophy*.

Gordon, J.-S., & Nyholm, S. (2022). Kantianism and the Problem of Child Sex Robots. *Journal of Applied Philosophy, 39*(1), 132–147.

Gunkel, D. (2012). *The Machine Question. Critical Perspectives on AI, Robots, and Ethics*. MIT Press.

Gunkel, D. (2018). *Robot Rights*. MIT Press.

Kurzweil, R. (2005). *The Singularity Is Near*. Penguin Books.

Loh, J. (2019). *Roboterethik. Eine Einführung*. Suhrkamp.

Ludwig, S. (2015). Credit Scores in America Perpetuate Racial Injustice: Here's How. *The Guardian*, October 13. https://www.theguardian.com/commentisfree/2015/oct/13/your-credit-score-is-racist-heres-why

Misselhorn, C. (2018). *Grundfragen der Maschinenethik*. Reclam.

Müller, V. C. (2020). Ethics of Artificial Intelligence and Robotics. *Stanford Encyclopedia of Philosophy*. https://plato.stanford.edu/entries/ethics-ai/

Müller, V. C., & Bostrom, N. (2016). Future Progress in Artificial Intelligence: A Survey of Expert Opinion. In V. Müller (Ed.), *Fundamental Issues of Artificial Intelligence* (pp. 553–571). Springer.

Nyholm, S. (2018a). The Ethics of Crashes with Self-Driving Cars: A Roadmap, I. *Philosophy Compass, 13*(7), e12507.

Nyholm, S. (2018b). The Ethics of Crashes with Self-Driving Cars, A Roadmap, II. *Philosophy Compass, 13*(7), e12506.

Nyholm, S. (2020). *Humans and Robots: Ethics, Agency, and Anthropomorphism*. Rowman and Littlefield.

Ord, T. (2020). *The Precipice: Existential Risk and the Future of Humanity*. Hachette Books.

Searle, J. R. (1980). Minds, Brains, and Programs. *Behavioural and Brain Sciences, 3*(3), 417–457.

Springer, A., Garcia-Gathright, J., and Cramer, H. (2018). Assessing and Addressing Algorithmic Bias – But Before We Get There. In *2018 AAAI Spring Symposium Series*, 450–454. https://www.aaai.org/ocs/index.php/SSS/SSS18/paper/viewPaper/17542

Vinge, V. (1993). The Coming Technological Singularity. How to Survive in the Post-Human Era. *Whole Earth Review*, Winter.

CHAPTER 3

What Are Human Rights?

Abstract This chapter presents a concise overview of human rights, exploring the distinctions between moral and legal human rights. It discusses the implications of human rights legislation in the context of AI, which pervades various aspects of human life, including healthcare, warfare, surveillance, governance, and autonomous transportation. AI's dual nature poses both benefits and risks to individuals and society, with potential misuse for immoral purposes or perpetuation of injustices such as machine bias. The chapter emphasises the need to evaluate AI's use within the framework of existing human rights legislation and consider revisions if necessary.

Keywords Human rights • Moral and legal human rights • Human rights legislation • Personhood

3.1 INTRODUCTION

This chapter provides a brief overview of the nature of human rights. Furthermore, it offers some initial insights regarding the differences between moral and legal human rights. It specifically considers human rights legislation and the practical implications of

J.-S. Gordon, *The Impact of Artificial Intelligence on Human Rights Legislation*, https://doi.org/10.1007/978-3-031-31388-2_3

using human rights as a universal *lingua franca* for moral reasoning and decision-making concerning complex cases such as the widespread use of AI in many different areas of human life (e.g. health care, warfare, surveillance, governance, autonomous transportation, automation, guidance, and support systems).

Like a double-edged sword used to defend and attack, AI can be used to do good, individually and for society at large, or to do wrong, such as if it were instrumentalised for immoral individual gains or to perpetuate injustices like machine bias. Therefore, we need to assess the opportunities and risks concerning the use of AI against the background of current human rights legislation and to revise existing legislation if changes are required. I agree with Raso et al. (2018, p. 8), who claim:

> There is a considerable value in adopting a human rights perspective to evaluating and addressing the complex impacts of AI on society. The value lies in the ability of human rights to provide an agreed set of norms for assessing and addressing the impacts of the many applications of this technology, while also providing a shared language and global infrastructure around which different stakeholders can engage.

The following sections will provide the basics to enable a better understanding of the issues at hand, especially concerning the impact of AI on human rights legislation (to be discussed in Chap. 4) and moving towards a Universal Convention for the Rights of AI Systems (Chap. 5).

3.2   What Are Human Rights?

3.2.1   *The Concept of Human Rights*

Human rights are primarily universal moral rights that set a minimum standard, valid independently of any state or legal recognition (Gordon, 2016). This is the standard view concerning moral human rights. Brian Orend (2002, p. 34) summarises the concept of human rights as follows:

A human right is a high-priority claim, or authoritative entitlement, justified by sufficient reasons, to a set of objects that are owed to each human person as a matter of minimally decent treatment. Such objects include vitally needed material goods, personal freedoms, and secure protections. In general, the objects of human rights are those fundamental benefits that every human being can reasonably claim from other people, and from social institutions, as a matter of justice. Failing to provide such benefits, or acting to take away such benefits, counts as rights violation. The violation of human rights is a vicious and ugly phenomenon indeed; and it is something we have overriding reasons to resist and, ultimately, to remedy.

In contrast, legal human rights depend significantly on the recognition of and commitment by member states to enforcing human rights in their domain (or in non-member states). In cases of great injustice, such as ethnic cleansing committed in non-member states, the political pressure on individual non-member states to solve the problems can be so substantial that many eventually defer to the international human rights community so as to avoid becoming subject to forms of socio-economic pressure (e.g. import and trade embargos).

The International Bill of Rights, a term used to denote the three most important human rights documents, includes the Universal Declaration of Human Rights (1948), which contains the general principles of human rights; the International Covenant on Civil and Political Rights (1966), which defines specific rights and their limitations in its particular domain; and the International Covenant on Economic, Social and Cultural Rights (1966), which also establishes particular rights and their limits in these fields.

Even though the documents of the International Bill of Rights are not, strictly speaking, legally binding (though some of the norms contained in it are), they have nonetheless become the most authoritative source for national and international legislation on issues concerning fundamental rights. For example, many of the rights enumerated in the Universal Declaration "are now norms of international customary law" (Buchanan, 2017, p. 7). It is fair to say that the United Nations significantly promotes the human rights agenda internationally, that various international tribunals apply international human rights law (e.g. in cases of genocides), and that judges of the supreme courts of member states commonly adhere to international legal human rights as one of the most significant legal sources along with their national constitutions. This development shows that the human rights agenda has become more and more legally binding as many of the enlisted rights are enforced on different levels within and between states.

One of the vital issues in the philosophy of human rights concerns the problem of their origin. Some scholars believe that human beings did not *invent* human rights but, on the contrary, *discovered* them by, for example, using reason or adhering to religious beliefs (i.e. natural law theory). Others, however, deny moral realism (the view that there exist moral facts independent from human existence) and hence argue that human rights, as part of *human* morality, are invented by us to regulate immoral and illegal actions committed either by states or by individuals.

Although the problem of origin is essential with regard to moral human rights and their justification, the domain of legal human rights seems somewhat detached from this particular problem. The reason is that international human rights legislation or human rights practice is based on international law and its legal infrastructure, which does not necessarily rely on questions of origin (unlike the moral domain of human rights).

3.2.2   The Main Features of Human Rights

Universality
Human rights are *universal* by nature and are not limited to a particular region. *All* human beings have them simply by being a member of the *human* race. This is an important feature. However, we will see later that this general idea raises some substantial issues by excluding non-human intelligent entities such as the great apes and dolphins, which many experts see as potential persons (Cavalieri,

2001; Francione, 2009). Many scholars believe being a person is key to having moral status and moral and legal rights (Kamm, 2007; Singer, 2009).

Highest Priority

Human rights enjoy the *highest priority*. This assertion is not about claiming champagne and caviar for all human beings but the essential things in life, such as food, shelter, and clean water, as well as security and liberty (i.e. the rights to life, liberty, and security). Fundamental human rights protect against substantial threats to human flourishing and well-being, primarily threats posed by the state.

Moral and Legal Human Rights

Furthermore, human rights can be distinguished according to their two different domains. The first domain concerns *moral human rights* while the second domain is about *legal human rights*. This distinction is significant because moral human rights are valid independently of state recognition. In contrast, legal human rights necessarily require a shared global legal infrastructure and understanding between different stakeholders regarding international human rights legislation, its legal enforcement, and practice. I will provide some brief remarks on the relationship between the two domains in Sect. 3.3.

Negative and Positive Human Rights?

In addition, even though the traditional distinction between negative and positive human rights has been extensively discussed in the past (Cranston, 1967), many *contemporary* human rights scholars believe that this distinction is nowadays of little value given the current human rights practice.

In former times, people believed that *negative* human rights, such as the right to equality before the law and the prohibition of slavery, torture, and inhumane treatment, had a kind of predominance among the different human rights. These rights were part of the so-called first generation of human rights (during the eighteenth century), which contained rights that were considered negative defence rights against the state. They were called "negative" because the state only had to *avoid* particular actions so as not to violate human rights (e.g. it could not infringe on the right to property, religious freedom, or free speech). Furthermore, these rights were considered "inexpensive" since the state did not have to invest anything (or at most only a small amount) to put them into practice and maintain their existence. Whether they have actually been inexpensive is a matter of debate, given the national and international legal infrastructure established to support and maintain proper human rights legislation.

In contrast, the second and third generations of human rights are considered *positive* human rights. The second generation of human rights (twentieth century) set forth a substantial catalogue of social and cultural human rights, such as the right to social security, the right to education, and the right to participate in the cultural life of one's society. The third generation of human rights (ca. 1980–2000) concerns group rights (especially rights of indigenous minorities), the right to national self-determination, and the right to a clean environment. Both of these two generations of human rights require substantial funds to put them into practice for human flourishing.

The fourth generation of human rights (latest developments) attempts to identify particular human rights concerns for either specific groups of people (e.g. in the context of disability rights) or the environment. Against this background, more and more scholars working in AI believe that it might be reasonable to start talking about "robot rights" (Gunkel, 2018) and to provide sophisticated, intelligent systems—especially human-like robots, once they emerge—with fundamental rights (Gordon, 2022; Gordon & Gunkel, 2022). The human rights of the fourth generation usually involve high implementation costs.

Some scholars claim that socio-economic rights are not "real" human rights because they are (again financially speaking) too demanding and go beyond the more *aspirational* character of human rights. Whether this is a good enough reason to argue against them is a matter of ongoing debate.

3.2.3   Different Types of Human Rights—A Classification

There is a common understanding that the corpus of human rights can be distinguished according to particular functions of human rights. As far as I can see, there are at least seven essential groups: security rights, due process rights, liberty rights, political rights, equality rights, social welfare rights, and group rights (e.g. Nickel, 2019: section 3). I will briefly comment on each group below.

Security rights protect people against crimes such as murder, massacre, torture, and rape by giving them substantial weight in human rights legislation. Especially when these actions are committed by the state (or a part of the state), the international community is asked to restore justice and prosecute all violations committed by state actors.

Due process rights protect people against abuses in the legal system, such as imprisonment without trial, secret trials, and excessive punishments. States which fail to meet these standards severely violate human rights.

Liberty rights protect freedoms in realms such as belief, expression, association, assembly, and movement. People's lives are essentially restricted if the state does not guarantee the rights mentioned. Enabling people to speak freely and discuss ideas with fellow citizens in public is one of the most fundamental democratic achievements in modern times.

Political rights protect liberty in politics, expressed through such actions as communicating, assembling, protesting, voting, and serving in public office. These political participation rights are the foundation of the modern state. Whether the state must necessarily be democratic by nature is a matter of debate.

Equality rights guarantee equal citizenship, equality before the law, and non-discrimination. Citizens whose rights are curtailed are victims of severe human rights violations. Prominent examples include African American citizens under racial segregation in the

US, black people during apartheid in South Africa, and most people under the Indian caste system (which is no longer officially in place but remains practically influential).

Social welfare rights require providing education to all people (especially children) and protection against severe poverty and starvation. It is usually impossible for people to fully enjoy other fundamental human rights if they are severely hungry or subject to impoverished living and health conditions that undermine human flourishing and well-being.

Group rights include the protection of ethnic groups against genocide and the confiscation of their territories and resources (e.g. the situation of the indigenous people in the US and Canada, the Aborigines in Australia, and the Jews in Nazi Germany). Overall, the development of the human rights practice to include group rights seems reasonable. Even though the status of group rights among human rights has been questioned and robustly discussed in the past, it nonetheless follows the inner logic of the nature of human rights as applied to groups rather than to individuals, as one might stress in their defence.

3.3   Moral and Legal Human Rights

There is a significant difference between moral human rights and (international) legal human rights (Buchanan, 2017; Wellman, 2005), even though most philosophers believe—perhaps rightly—that moral human rights *precede* legal human rights and that the latter category could be seen as legally realising the former. This standard view or "mirroring view" (Buchanan, 2017, p. 21) is held by many prominent human rights scholars, such as John Tasioulas (2002), Joseph Raz (2010), James Griffin (2009), and Johannes Morsink (1999, 2009). Prominent opponents include Allen Buchanan (2017) and Charles Beitz (2009).

Until very recently, Buchanan agreed with the standard view; however, in his recent book *The Heart of Human Rights* (2017), he criticises what he calls "the mirroring view" and offers a framework that could justify international legal human rights independently from moral human rights. Even though he agrees with Wellman that "the founders of the international legal human rights system thought that the rights it contains presuppose corresponding moral rights", he adds that "from that it does not follow that the best reconstruction of the system will retain that assumption" (2017, p. 4, footnote 1).

Buchanan acknowledges that the "justificatory rhetoric" contained in the Universal Declaration and the two Covenants suggests that "the function of the documents is to specify international legal analogs of a set of preexisting moral human rights" (2017, p. 21). According to Buchanan, however, this cannot mask the fact that there is no consensus on the nature of moral human rights. Therefore, it would not make sense to realise them legally in the first place (because there is no common understanding of what they are). Moral human rights are neither "necessary" nor "sufficient" to justify legal human rights (Buchanan, 2017, p. 21, 43). In contrast, he argues that one must adhere to a "pluralistic justificatory methodology" which does not appeal to "preexisting moral rights, but instead … to other moral considerations" (Buchanan, 2017, p. 43). The two main functions of international legal human rights, according to Buchanan, are (1) to limit state sovereignty in cases of abusive behaviour by the state and (2) to affirm and protect the "equal basic status of all individuals"[1] and to promote human well-being so that people can live a minimally good or decent life. I think Buchanan is right when he claims the following:

> For in the Practice, it is international legal human rights that serve to limit sovereignty, not the right of any "folk" or philosophical theory of moral human rights. No folk or philosophical theory of moral human rights is determinate enough or widely accepted enough to play the roles in the Practice that international legal human rights play. (Buchanan, 2017, p. 15)

[1] Buchanan (2017, p. 37).

Even though the above statement is correct, it does not follow from this that moral human rights are less significant or do not provide the basis for international legal human rights. Legal human rights have a more substantial impact concerning their enforcement *as* international codified law, which differentiates them from even the most basic moral human right unless it has been put into law. Legal enforceability is the crucial difference. Furthermore, there is, in general, no one-to-one relationship between moral and legal human rights. The idea that a particular legal human right in the Universal Declaration *always* has a specific moral counterpart is misleading. However, these considerations do not undermine the idea that moral human rights are the foundation of legal human rights.

The idea that the justification of international legal human rights should be independent of moral human rights is far-fetched, given that the International Bill of Rights was initially meant to be the legal instantiation of moral human rights in the first place. This general view does not depend on a strict causal relationship between moral and legal human rights and certainly not on the establishment of one-to-one relationships. But what are the alternative options? There are at least four different ways in which one could perceive the relationship between moral and legal human rights:

1. One-to-one relationship: Each legal human right can be traced back to one moral human right.
2. One-to-many relationship: One legal human right can be traced back to multiple moral human rights.
3. Many-to-one relationship: Multiple legal human rights can be traced back to one moral human right.
4. Many-to-many relationship: Various legal human rights can be traced back to multiple moral human rights.

The most convincing view is that different legal human rights can be often traced back to different moral human rights (a many-to-many relationship), even though one cannot rule out the possibility that one legal human right could be an instantiation of exactly one moral human right (see the example below). The relationship between moral and legal human rights is complex. But that does not mean that one cannot show that the latter is the realisation of the former, as Buchanan claims in arguing that his justification of international legal human rights depends not on moral rights but on "other moral considerations".

For example, the legal human right *to be free from slavery* seems to have its justification in the moral human right *to be free from slavery*. The reason why we think that slavery is immoral provides the reasoning for adopting the moral human right to be free from slavery. In other words, since we believe that autonomy and the idea of making our own decisions based on our preferences and life plans are of great significance for any free and equal person, we conclude that this seems to be something that should be universally protected, morally speaking. The highest degree of moral protection is framed in the language of moral human rights. It appears that Buchanan directly jumps to the *general reasoning* ("other moral considerations") by leaving out the moral human rights talk, which sits in the middle between legal human rights on one hand and the moral justification of the moral human right on the other hand.

Buchanan's justification for international human rights legislation strongly relies on the *empirical* or practical[2] dimensions of human rights and on an *institutional* foundation of human rights based on the United Nations system of bodies and courts. His avoidance of justifying legal human rights by appealing to moral human rights could be an advantage, because otherwise the whole weight of justification concerns the reasonableness of a theory of moral human rights. If such a theory stands on shaky ground, the justification for legal human rights could also be questioned. In that respect, Buchanan might be correct in offering an institutional foundation for the human rights legislation that is independent of moral human rights.

---

[2] Buchanan (2017, pp. 5–6).

My primary aims in the book are to examine AI's impact on human rights (legislation) and to provide a moral justification for creating a *Universal Convention for the Rights of AI Systems.* This so-called AI Convention can be seen as the legal instantiation of the moral human rights agenda in the context of AI systems. The following chapters will depict this line of reasoning.

3.4   Human Rights and Personhood

Many scholars believe that the concept of human dignity is the foundation for the ascription of human rights. Human beings have human rights by virtue of their humanity. But what does that mean? There are at least two main responses to this question. One could argue either that only beings who share human DNA have human rights (a biological argument) or that only people with human dignity are entitled to human rights (a religious or naturalistic argument). The two lines of argumentation are as follows.

3.4.1   The Human DNA Argument

1. Only human beings are entitled to human rights.
2. A being is human when it has human DNA.
3. Only beings with human DNA are entitled to human rights.

This first argument rules out the possibility that non-human beings are entitled to human rights because the biological feature of *being human* is part of the definition. That means intelligent aliens, smart animals such as the great apes and dolphins, and superintelligent robots could never be entitled to "human rights"—even if they were more intelligent than human beings—because they are essentially not human by nature. Peter Singer (2009) famously claims that this type of (biological) reasoning is prone to speciesism, that is, racism on the level of species, and hence should be rejected. Being human is neither necessary nor sufficient for entitlement to fundamental rights.

3.4.2 *The Human Dignity Argument*

1. Only human beings have dignity.
2. A being with dignity has a full moral status.
3. A full moral status guarantees moral personhood.
4. Moral personhood provides for moral and legal rights (including human rights).

This second argument can be understood in either *religious* or *naturalistic* ways. Regarding the religious interpretation, proponents could argue that human beings have dignity because they are created in the *imago dei* (the image of God) and hence are entitled to human rights. Or, in the naturalistic interpretation, "nature" has endowed human beings with dignity by virtue of their humanity, uniqueness, or intelligence. Both proposals suffer from a lack of adequate supportive reasoning.

Critics could object that we do not know whether God exists (since the current proofs of the existence of God have all failed) and that the idea of a conscious nature with its own will seems highly questionable because it is unlike anything else we have discovered in the natural sciences so far. Furthermore, scholars such as Ruth Macklin (2003) have claimed that the idea of dignity is too *vague* and *abstract* to be of any use for ethical reasoning and decision-making. If I understand her correctly, that also includes using it to justify human rights.

A more promising line of argument uses the concept of personhood to justify a being's entitlement to human rights. What does that mean? It means that human beings are entitled to human rights because they are *persons*. Persons have moral and legal status and hence—based on this status—have moral and legal rights. But what makes a person a person? Some scholars, such as utilitarian authors, believe that personhood should be defined by sentience (the ability to feel pain). In contrast, Kantians claim that personhood should best be identified in terms of sapience (the ability to reason). Or, according to the proponents of the relational approach

(Coeckelbergh, 2014; Gunkel, 2012), personhood should be defined by a being's social relations with other beings, independent of whether a being feels pain or can reason. In *Artificial Moral and Legal Personhood* (2021), I have argued that personhood should be defined by several morally relevant criteria, such as rationality, intelligence, autonomy, self-awareness, and the ability to make moral decisions. Furthermore, these criteria should be species-independent and applicable to humans and non-human beings, so as to avoid the objection of speciesism.

Not all human beings are persons and therefore protected by moral rights (Singer, 2009). Conversely, it could be the case that some non-human persons exist who might be entitled to moral and legal rights or even to human rights (Cavalieri, 2001; Francione, 2009; Gordon, 2022; Gordon & Pasvenskiené, 2021). The question remains whether one should use the concept of *human* rights for non-human persons, since non-human persons are, obviously, not human. Since human rights conceptually concern humans, it might be reasonable to select a different notion, such as "fundamental rights", to cover non-human persons like animals and artificial machines (see Chap. 5).

3.5   Conclusions

This chapter has offered a brief overview of the nature of human rights, without delving deeply into the many details of this complex concept. The information presented above is sufficient to understand the concept in general terms and to demonstrate the difficulties related to the concept that will be discussed in the following chapters, especially when the impact of AI on human rights is examined.

References

Beitz, C. (2009). *The Idea of Human Rights*. Oxford University Press.
Buchanan, A. (2017). *The Heart of Human Rights*. Oxford University Press.
Cavalieri, P. (2001). *The Animal Question: Why Non-Human Animals Deserve Human Rights*. Oxford University Press.

Coeckelbergh, M. (2014). The Moral Standing of Machines: Towards a Relational and Non-Cartesian Moral Hermeneutics. *Philosophy & Technology*, *27*(1), 61–77.

Cranston, M. (1967). Human Rights, Real and Supposed. In D. D. Raphael (Ed.), *Political Theory and the Rights of Man* (pp. 43–53). Indiana University Press.

Francione, G. L. (2009). *Animals as Persons: Essays on the Abolition of Animal Exploitation*. Columbia University Press.

Gordon, J.-S. (2016). Human Rights. *Oxford Bibliographies in Philosophy* (Ed. Duncan Pritchard), Published Online.

Gordon, J.-S. (2021). Artificial Moral and Legal Personhood. *AI & Society*, *36*(2), 457–471.

Gordon, J.-S. (2022). Are Superintelligent Robots Entitled to Human Rights? *Ratio*, *35*(3), 181–193.

Gordon, J.-S., & Gunkel, D. (2022). Moral Status and Intelligent Robots. *Southern Journal of Philosophy*, *60*(1), 88–117.

Gordon, J.-S., & Pasvenskiené, A. (2021). Human Rights for Robots? A Literature Review. *AI and Ethics*, *1*, 579–591.

Griffin, J. (2009). *On Human Rights*. Oxford University Press.

Gunkel, D. (2012). *The Machine Question: Critical Perspectives on AI, Robots, and Ethics*. MIT Press.

Gunkel, D. (2018). *Robot Rights*. MIT Press.

Kamm, F. (2007). *Intricate Ethics: Rights, Responsibilities, and Permissible Harm*. Oxford University Press.

Macklin, R. (2003). Dignity is a Useless Concept: It Means no More Than Respect for Persons or Their Autonomy [Editorial]. *BMJ: British Medical Journal*, *327*(7429), 1419–1420. https://doi.org/10.1136/bmj.327.7429.1419

Morsink, J. (1999). *Universal Declaration of Human Rights: Origins, Drafting, and Intent*. University of Pennsylvania Press.

Morsink, J. (2009). *Inherent Human Rights: Philosophical Roots of the Universal Declaration*. University of Pennsylvania Press.

Nickel, J. (2019). Human Rights. *Stanford Encyclopedia of Philosophy*, Published Online. https://plato.stanford.edu/entries/rights-human/

Orend, B. (2002). *Human Rights: Concept and Context*. Broad View Press.

Raso, F., Hilligoss, H., Krishnamurthy, V., Bavitz, C., & Levin, K. (2018). *Artificial Intelligence and Human Rights: Opportunities and Risks*. Berkman Klein Center for Internet and Society Research Publication. http://nrs.harvard.edu/urn-3:HUL.InstRepos:38021439

Raz, J. (2010). Human Rights Without Foundations. In S. Besson & J. Tasioulas (Eds.), *The Philosophy of International Law* (pp. 321–337). Oxford University Press.

Singer, P. (2009). Speciesism and Moral Status. *Metaphilosophy*, *40*(3–4), 567–581.

Tasioulas, J. (2002). Human Rights, Universality, and the Values of Personhood: Retracting Griffin's Steps. *European Journal of Philosophy.*, *10*(1), 79–100.

Wellman, C. (2005). *The Moral Dimension of Human Rights*. Oxford University Press.

The Impact of Artificial Intelligence on Human Rights

Abstract This chapter synthesises the content from previous chapters to assess the impact of AI on human rights. It begins with a brief overview of the AI–human rights relationship, followed by an in-depth exploration of two conflicting perspectives. The first perspective addresses human rights infringements by AI systems, while the second, less-studied perspective investigates the violation of "machine rights". The chapter concludes by outlining potential solutions for the issues discussed.

Keywords Human rights and AI • Zones of conflict • Impact of AI • Human rights legislation • Universal Declaration of Human Rights • Machine rights

4.1 INTRODUCTION

This chapter brings together the points mentioned in the previous chapters so as to determine the impact of AI on human rights. The next section offers a brief overview of the relationship between AI and human rights as a starting point; after that, I examine the so-called zones of conflict in greater detail by fleshing out two different perspectives. The first perspective examines human rights

J.-S. Gordon, *The Impact of Artificial Intelligence on Human Rights Legislation*, https://doi.org/10.1007/978-3-031-31388-2_4

infringements by AI systems, while the second perspective—on which scholars have rarely focused—determines the violations of "machine rights". Both perspectives are analysed in the same three time frames delineated previously: short-term (early twenty-first century), mid-term (2040–2100), and long-term (starting from the 2100s). The chapter ends with some brief conclusions on how to solve the issues raised here, foreshadowing the fifth chapter, which proposes moving towards a *Universal Convention for the Rights of AI Systems.*

4.2   Artificial Intelligence and Human Rights— An Overview

The tremendous technological developments concerning AI, robotics, computer science, informatics, and related fields have already produced substantial socio-political, moral, and legal implications of utmost significance (Gordon & Nyholm, 2021; Loh, 2019; Misselhorn, 2018; Müller, 2020). Efforts to address issues of autonomous transportation (problems in cases of lethal accidents), autonomous drones in warfare (should machines make the final decisions?), AI use in parole decisions in the US (which revealed the problem of machine bias), AI-based analysis of job applications in the private sector (which revealed gender bias and injustice towards certain ethnic groups), and AI use in the context of loan applications (which exhibited biases by ethnicity and social class) have already demonstrated the significant challenges we are facing as AI increasingly pervades our world (Gordon & Nyholm, 2021).

The examples mentioned above are only the tip of the iceberg and call for a thorough analysis of the impact of AI on human rights legislation. Suppose, for example, that human beings are disadvantaged by an application of AI, such as in a loan application or a parole-related risk assessment. This could constitute a severe violation of a person's human rights (Bacciarelli et al., 2018). Current research concerning AI has effectively identified many normative and legal issues (Anderson & Anderson, 2011; Lin et al., 2014; Wallach & Allen, 2010); however, we are only at the beginning in terms of taking broader implications for international human rights legislation into account. In particular, only a few publications have examined the impact of AI on human rights, a topic to which I have contributed in the past (Gordon, 2022; Gordon & Pasvenskiene, 2021; Livingston & Risse, 2019; Risse, 2019).

Human rights are, first and foremost, fundamental moral rights of the highest significance and, secondarily, universal legal rights of utmost importance, as agreed upon in the 1948 Universal Declaration of Human Rights (Gordon, 2016). When this document was composed, the topic of AI and its impact on ethics and law was, of course, of no concern whatsoever. This situation has changed dramatically within the last few decades and has become highly worrisome in the last few years. We are now compelled to think more deeply and extensively about the relationship between AI and human rights, particularly the impact of AI on human rights legislation (Bacciarelli et al., 2018; Livingston & Risse, 2019; Raso et al., 2018; Risse, 2019).

There are at least two main perspectives on this topic. The first perspective concerns human rights violations committed by autonomous artificial systems that infringe upon human beings' fundamental moral and legal rights (parole decisions, loan applications, job applications, etc.). The second perspective, which has been rarely taken into account thus far, concerns violations of *machine rights* or *robot rights* (Gordon & Gunkel, 2022; Gunkel, 2012, 2018). Here, the basic idea is that if autonomous and intelligent machines become as capable as human beings or even supersede their cognitive abilities, they should be entitled to fundamental moral and legal rights based on their high moral and legal status (Gordon, 2020a).

Both perspectives should be evaluated against the background of the different time frames and the relevant topics that emerge in AI. Gordon and Nyholm (2021) mentioned some of the critical issues that humanity will (most likely) face in view of the pervading influence of AI in our daily lives:

1. Short-term (early twenty-first century): autonomous systems (transportation, weapons), machine bias in law, privacy and surveillance, the black box problem, and AI decision-making

2. Mid-term (from the 2040s to the end of the century): AI governance, confirming the moral and legal status of intelligent machines (artificial moral agents), human–machine interaction, and mass automation
3. Long-term (starting with the 2100s): technological singularity, mass unemployment, and space colonisation

This book seeks to determine possible issues surrounding AI's impact on human rights (legislation), review their significance with regard to human rights violations, and provide some initial answers on how to deal with particular situations. I maintain that we have already seen several critical, emerging issues that should prompt us to revise our human rights legislation with regard to the first perspective, or human rights violations resulting from the new era of AI. The second perspective already requires substantial research even though today's technological achievements are still far from the strong AI required for artificially intelligent machines that match or supersede human capabilities (Bostrom, 2014; Müller & Bostrom, 2016).

As noted above, one should prepare in advance for the advent of highly sophisticated machines so as to deal with the far-reaching ethical and legal implications of their existence before it is too late (Bostrom, 2014; Ord, 2020). Current discussions are considering whether intelligent machines should be entitled to moral and legal personhood and the related moral and legal rights once they reach a certain level of sophistication (Gordon, 2021; Gordon & Gunkel, 2022). Some argue that full moral status necessarily requires being human. In contrast, others contend that idiosyncratic features such as sharing human DNA should not play any role in an entity's entitlement to moral and legal rights.

The impact of AI on human rights legislation is a genuine concern, not something akin to fairy tales (Livingston & Risse, 2019; Risse, 2019). This is why one should start working on strategies to solve any upcoming issues and become prepared for both the present and the future. We have already seen that AI significantly impacts our daily lives and pervades human activities. We cannot put the genie back in the bottle, but we need to examine and thoroughly evaluate the advantages and disadvantages of emerging technologies for human society. I want to contribute to this endeavour by systematically examining the impact of AI on the Universal Declaration of Human Rights (1948), going through the different rights enumerated therein to provide a technology assessment concerning short-term, mid-term, and long-term scenarios. This will give us a solid foundation for developing a *Universal Convention for the Rights of AI Systems.*

4.3 Universal Declaration of Human Rights— Abridged Version

The following[1] is a list of the rights covered in the Universal Declaration of Human Rights (1948), which we can consult to identify what rights are violated in cases of infringements by AI:

Article 1	Right to equality
Article 2	Freedom from discrimination
Article 3	Right to life, liberty, and personal security
Article 4	Freedom from slavery
Article 5	Freedom from torture and degrading treatment
Article 6	Right to recognition as a person before the law
Article 7	Right to equality before the law
Article 8	Right to remedy by competent tribunal
Article 9	Freedom from arbitrary arrest and exile
Article 10	Right to fair public hearing
Article 11	Right to be considered innocent until proven guilty
Article 12	Freedom from interference with privacy, family, home, and correspondence
Article 13	Right to free movement in and out of the country
Article 14	Right to asylum in other countries from persecution
Article 15	Right to a nationality and the freedom to change it
Article 16	Right to marriage and family
Article 17	Right to own property
Article 18	Freedom of belief and religion
Article 19	Freedom of opinion and information
Article 20	Right of peaceful assembly and association

[1] The list of articles can be found at http://hrlibrary.umn.edu/edumat/hredu-series/hereandnow/Part-5/8_udhr-abbr.htm.

Article 21 Right to participate in government and in free elections
Article 22 Right to social security
Article 23 Right to desirable work and to join trade unions
Article 24 Right to rest and leisure
Article 25 Right to adequate living standard
Article 26 Right to education
Article 27 Right to participate in the cultural life of a community
Article 28 Right to a social order that articulates this document
Article 29 Community duties essential to free and full development
Article 30 Freedom from state or personal interference in the above rights

4.4   Zones of Conflict I—Machines Violate Human Rights

This section provides a table summarising the impact of AI on human rights, especially the first perspective (i.e. ways in which machines violate the human rights of humans), by identifying the infringed rights of the Universal Declaration of 1948 (Table 4.1).

4.4.1   Table

Table 4.1   Table of the short-term impact of AI on human rights (early twenty-first century)

Zones of Conflict I	Infringed human rights	Suggested corrections
Law: Criminal justice and risk assessment (machine bias)	Articles 1, 2, 7, 8, 10, 11, 30	Until machine bias cannot be ruled out, the use of software such as COMPAS in risk assessments for parole decisions must be avoided
Predictive policing, racial profiling (machine bias)	Articles 1, 2, 5, 7, 9, 11, 30	AI-assisted predictive policing and profiling based on machine bias must be avoided
Facial recognition systems in policing and governance	Articles 1, 2, 3, 5, 9, 11, 12, 19, 20, 30	AI-assisted facial recognition systems for policing (in case machine bias is involved) and widespread monitoring of the behaviour of citizens must be avoided

(*continued*)

Table 4.1 (continued)

Zones of Conflict I	Infringed human rights	Suggested corrections
Loan applications (machine bias)	Articles 1, 2, 5, 12, 17, 22, 25, 30	AI-assisted decisions about loan applications should not be used if machine bias cannot be ruled out
Human resources: Recruitment and hiring (machine bias)	Articles 1, 2, 5, 12, 22, 23, 25, 30	AI-assisted software should not be used in human resources if machine bias cannot be ruled out
Autonomous transportation	Article 3	AI-assisted software for self-driving vehicles which does not meet acceptable standards endangers people's lives and therefore should not be used
Natural language processing techniques: Automated translations and online content moderation (standards enforcement)	Articles 1, 2, 5, 19, 30	Natural language processing techniques can be the basis for discrimination when it comes to adequate translations (gender issues)
Healthcare: Diagnostics, supervision, and monitoring	Articles 12, 30	AI-assisted software might violate the right to privacy
Military (war robots, drones)	Article 3	The use of autonomous AI-assisted software in warfare could lead to civilian casualties
Education: Essay grading	Articles 1, 2, 26, 30	AI-assisted software in education could either promote the capabilities of minorities or undermine their right to an equal education
Freedom of expression	Articles 1, 2, 3, 9, 12, 18, 19, 20, 30	AI-assisted software can monitor and document the location and behaviour of people and thereby influence people's behaviour in negative ways (e.g. suppressing their opinions and participation in demonstrations)

4.4.2   Comments

The preceding, self-explanatory table shows that many areas of conflict exist between AI and human rights, especially concerning the right to equality (Article 1) and freedom from discrimination (Article 2). The significance of both rights in the context of AI has been highlighted in The Toronto Declaration (Bacciarelli et al., 2018). Many of the mentioned problems are caused by *machine bias*, where the decisions made by an AI system *maintain* and *aggravate* existing human bias due to unprepared historical data or defective software (or both), making it impossible for the victims to be heard (e.g. in law or loan applications).

Furthermore, the situation is exacerbated because in most cases, one does not know *how* the AI system makes its decisions. This is called the black box problem. The right to know the reasons for decisions, especially rejections of applications, requires that AI systems must become "explainable". At this moment, it is still uncertain whether we can build self-learning autonomous machines (with a deep neural network architecture) that clearly explain how they arrived at their decision.

However, to deal with machine bias, which (given its centrality in many applications and its substantial impact on many people around the world) is probably one of the most fundamental problems in the context of AI ethics, the following solutions have been suggested:[2]

1. The human-in-the-loop approach (humans should make the final decisions)
2. Including diverse team members of different ethnicities and genders
3. Inclusive training for teams at all stages of design and programming

---

[2] See also The Toronto Declaration (Bacciarelli et al., 2018) and Raso et al. (2018).

4. Avoiding the use of unprepared historical data
5. A proper testing phase to exclude issues related to machine bias
6. Monitoring and recording the results of the AI system once it
   is in use and reprogramming its software in case of malfunction

It is of utmost importance to make people, especially programmers, aware of the deeply rooted ethical problems concerning artificial, autonomous decision-making (Gordon, 2020b). Numerous
examples[3] document this fundamental problem. Undoubtedly, current AI systems violate numerous human rights, which is unaccept-
able and calls for an immediate action plan to avoid further
violations. On the other hand, it seems impossible to ask for a widespread hold on the development of AI systems, given their exten-
sive involvement in our work and spare time. The best we can do is
to determine the current problems, fix them, and do better with
upcoming AI systems. That would already be a substantial contri-
bution to avoid violating the rights to equality and to freedom from
discrimination.

Furthermore, some problems are not caused by AI systems but
occur because human beings use the systems in a particular way.
For example, widespread and permanent surveillance coupled
with a system that relates points to the social behaviour of citizens, as is currently done in China, substantially violates several
human rights, such as the right to life, liberty, and personal security (Article 3), freedom from interference with privacy, family,
home, and correspondence (Article 12), freedom of belief and
religion (Article 18), freedom of opinion and information (Article
19), the right of peaceful assembly and association (Article 20),
the right to desirable work and to join a trade union (Article 23),
the right to an adequate living standard (Article 25), the right to

---

[3] For example: Machine bias in law, in banking and finance systems, in predictive
policing and racial profiling, in loan applications, in recruitment and hiring, and in
facial recognition systems in policing and governance and more.

a social order that articulates this document (Article 28), and freedom from state or personal interference in the above rights (Article 30). Anti-social behaviour, that is, behaviour that the Chinese government defines in a particular way, could eventually lead to losing one's job, being denied a loan, and other social disadvantages. Comprehensive state surveillance violates citizens' right to privacy and urges them not to express any critical opinions, especially ones related to the state. Everything against the state is punished, and everything that supports the state's position is rewarded. This situation is possible only through the use—or, rather, misuse—of AI (Table 4.2).

4.4.3   Table

Table 4.2   Table of mid-term impact of AI on human rights (2040–2100)

Zones of Conflict I	Infringed human rights	Assessment
Machine bias in its various forms	Articles 1, 2, 7, 9, 11, 17, 22, 23, 25, 26, 30	Machine bias is part of many different applications in our world and must be avoided in all circumstances (see the suggestions in Sect. 4.4.2)
Mass automation: Many people will lose their jobs	Articles 17, 22, 23, 25, 30	If people lose their jobs to AI systems, one must consider providing them with financial compensation. People have the right to live (which includes the necessary finances for food, shelter, and housing)
AI governance	Article 21	A full-fledged system of AI governance might violate the human right to participate in governance at various levels of the state
Freedom of expression	Articles 18, 19, 20	The widespread permanent monitoring of people's social behaviour by AI systems undermines freedom of expression. Such situations must be avoided

4.4.4 Comments

Whether the problem of machine bias will be fully solved in the mid-term is a matter of debate. It will certainly take some time. The most important step at this point is to monitor and review the process of designing and programming AI systems and to examine the results once the AI system is working so as to take measures against any bad results.

Furthermore, the widespread use of AI systems in the working world will necessarily lead to a situation where many people will lose their jobs, either because machines do it better and more efficiently or because the jobs are too dangerous or dull. On one hand, this is positive since it helps people find more meaningful jobs (or use their free time for their family and hobbies); on the other hand, it can be a substantial threat since people need money to make a living, and if they have no employment, many will face significant problems. In addition, mass automation could cause people to experience existential boredom (Bloch, 1954). The topic of whether less work would be preferable to the current situation has been examined by Danaher (2019a).

The more advanced AI systems become, the more complex tasks they can handle, including tasks related to the organisation and management of everyday processes in companies and states. Some futurists yearn for the day when AI systems fully govern whole so-called smart cities and even states with little or no influence by people. Whether this is a utopia or a dystopia remains to be seen. But the most problematic issue concerns the lack of participation in governance by citizens in times of highly advanced AI systems.

The topic of freedom of expression remains critical, given the possibility of using AI for questionable means (as with China's social points system). A more powerful AI system could create a situation where people feel constantly monitored and act in ways that do not meet the standards for flourishing human life. To freely express one's opinions or demonstrate when one wants is an essential human activity that must be protected against oppression caused by a state or by AI itself (see the movie *iRobot*). We must be aware of the dangers that could undermine our ways of living and we must implement safeguards to avoid unpleasant consequences (Table 4.3).

4.4.5   Table

Table 4.3 Table of the long-term impact of AI on human rights (starting from the 2100s)

Zones of Conflict I	Infringed human rights	Assessment
Competition for global resources	Articles 1, 2, 3, 20, 22, 25	A fair distribution of global resources must be guaranteed
AI governance	Articles 1, 2, 3, 4, 6, 7, 9, 12, 13, 19, 20, 21, 28, 30	There should be a human-in-the-loop for final decisions to avoid any form of discrimination against humans
Mass unemployment	Articles 3, 17, 23, 25	In the case of mass unemployment, humans must be compensated financially
Mass surveillance	Articles 2, 12, 18, 19, 23, 30	Mass surveillance must be prohibited

4.4.6   Comments

The long-term impact of AI on human rights is based entirely on the prospects for further technological achievements in AI, computer science, robotics, and related fields. Many scholars believe that AI machines will develop a high level of rationality, autonomy, intelligence, and self-awareness. In particular, Kurzweil (2005) and Bostrom (2014) believe that we will see the emergence of singularity, superintelligence, and AGI. The age of singularity, the point at which AI systems become smarter than humans, can be considered a game changer. For example, should superintelligent robots come into existence, we will be faced with difficult moral problems concerning their moral and legal status and their entitlement to moral and legal rights (Gordon, 2020a; Gordon, 2022; Gordon & Pasvenskiene, 2021; Gunkel, 2018).

Balancing the fundamental rights of humans and superintelligent robots will undoubtedly be an essential discussion point, especially regarding the competition for global resources to ensure the survival of both species under limited conditions. The potential for clashes between the existential interests of humans and superintelligent robots will depend mainly on how the latter species acts, given the unmistakable power imbalance between humans and superintelligent machines. This is the situation Bostrom (2014), Ord (2020), Gordon (2022), and others warn us about, since humans will not be able to resist such machines, given their exceptional capabilities. That is why Bostrom and Ord argue that one must *align* the superintelligent machines' views with our moral values and norms so that they will eventually accept our ways of dealing with conflicts peacefully and come to an agreement that serves both parties.

It is quite possible that superintelligent AI systems will eventually organise *everything* in a nation-state (complete AI governance) and that human beings could enjoy whatever they want, regardless of whether they have jobs. However, some scholars argue that this situation could lead to "a crisis in moral agency" (Danaher, 2019b), human "enfeeblement" (Russell, 2019), or "de-skilling" in different fields of human life (Vallor, 2015, 2016). Complete AI governance, combined with smart cities (which will be the standard by then), could lead to a situation where all people's actions are permanently monitored and assessed according to specific criteria (mass surveillance). The vital question of who decides what criteria are used remains unresolved, and it is unclear whether humans will have a say in this decision. The right to privacy and freedom of expression could be under fire at that time. To avoid this unfortunate outcome, we must consider ways to preserve human freedom and remain active and autonomous.

4.5   Zones of Conflict II—Humans Violate "Machine Rights"

Whereas the aforementioned *zones of conflict* concern violations committed by AI systems, this section offers an overview of possible violations of so-called machine or robot rights caused by humans. The idea of machine rights, that is, the moral and legal nature of intelligent and autonomous machines, is of recent origin. Although most scholars are still hesitant to accept the idea of morally relevant, species-independent criteria for being entitled to moral and legal rights, others, such as Gordon (2020a, 2021, 2022), Gordon and Gunkel (2022), and Gunkel (2018), believe we should be open to the view that highly advanced intelligent machines are *entitled* to fundamental rights too, regardless of what human beings think about the matter (Table 4.4).

4.5.1   Table

Table 4.4   Table of violations of machine rights

Zones of Conflict II	Infringed machine rights	Assessment
Short-term (early twenty-first century)		
Bodily integrity	None	This might be a problem for the "brutalisation" of the human character when *social robots* are involved (e.g. Darling, 2016)
Bodily destruction	None	This might be a problem for the "brutalisation" of the human character when *social robots* are involved (e.g. Darling, 2016)
Creative work: Painting, music	None	None
Health sector: Discoveries, diagnostics	None	None
Mid-term (2040–2100)[a]		
Bodily integrity	None	This might be a problem related to the "brutalisation" of the human character when *social robots* are involved (e.g. Darling, 2016)
Bodily destruction	None	This might be a problem related to the "brutalisation" of the human character when *social robots* are involved (e.g. Darling, 2016)

(*continued*)

Table 4.4 (continued)

Zones of Conflict II	Infringed machine rights	Assessment
Forced labour	None	The more human-like and "alive" machines or robots are, the more difficult it will be to use them as mere tools
Sex robots, companions	None	The more human-like and "alive" machines or robots are, the more difficult it will be to use them as mere tools
Creative work: Painting, music, writing	None	Even though the AI system is not (fully) conscious yet, the question of who "owns" the product or output could be raised
Strategy development (e.g. in companies)	None	Even though the AI system is not (fully) conscious yet, the question of whose responsibility is at stake must be examined carefully
Long-term (starting from the 2100s)[b]		
Bodily integrity	Articles 1, 2, 3, 5, 18, 19, 30	AI systems' right to bodily integrity must be ensured
Bodily destruction	Articles 1, 2, 3, 4, 5, 30	It is not permissible to destroy an AI machine without an exceptionally good reason
Forced labour	Articles 1, 2, 3, 4, 17, 20, 22, 23, 24, 30	No forced labour is legitimate
Sex robots, companions	Articles 1, 2, 3, 4, 5, 17, 20, 22, 23, 24, 30	No forced sex labour or companionships are legitimate
Creative work: Painting, music, writing	Articles 1, 2, 4, 17, 20, 22, 23, 24, 30	Must be treated in similar ways to human creative work
Strategy development (e.g. in companies)	Articles 1, 2, 4, 17, 20, 22, 23, 24, 30	Must be treated in similar ways to human creative work

[a]This portion of the table presupposes that AGI and superintelligence have not emerged yet
[b]This portion of the table presupposes that AGI and/or superintelligence have emerged

4.5.2   Comments

Violations of machine rights come into play once the AI systems reach a minimum level of rationality, autonomy, intelligence, and self-awareness. Many current AI systems already have a high level of autonomy, but this is not enough to justify their entitlement to strong moral and legal rights. Such systems must also be rational, intelligent, and self-aware. What level they must attain to meet that standard remains a matter of debate. In other words, the zones of conflict during the early twenty-first century are rather limited or almost non-existent, given the limited sophistication of current AI systems. This does not mean that current machines do not perform well enough; on the contrary, existing AI systems (e.g. self-learning, neural networks) are already capable of great things, often surpassing human capabilities (e.g. diagnostics in healthcare). But they are not currently at a level where their rights could be violated.

However, one might still ask who deserves the credit for current machines' ability to solve complex issues, develop successful business strategies, produce extraordinary pieces of art (including both painting and music), and make significant discoveries (e.g. in healthcare). The programmers built the architecture and provided the system with the initial capabilities, but the AI system, if it is self-learning, is the decisive ingredient responsible for all subsequent actions and results. That's why it is essential to implement ethics in AI systems so that they will learn the difference between morally good and bad behaviour right from the start. Eliminating cancer should not be achieved by killing all human beings who might be possible hosts of cancer, but by actually finding a remedy for the disease. Machines must be able to understand the difference between the two options.

The more compelling cases concern mid-term and long-term issues once AI systems have reached a level of understanding that is comparable to or surpasses human capabilities. The problem of forced labour (regardless of profession) and especially sex robots (including companion robots) will become of utmost significance (e.g. Gordon & Nyholm, 2022; Levy, 2008; Nyholm & Frank, 2017, 2019). Another challenging problem concerns whether highly advanced machines should be allowed to make their own decisions, which might include deciding to quit the job and do something else. Admittedly, this will not happen within the next few decades, but once AGI exists, this issue could become relevant within a few years.

At that time, humanity will be at the crossroads of either allowing such machines to follow their chosen path or trying to prevent them from doing so. The latter would be a case of "modern machine slavery". But the issue is more complicated. If we are confronted by superintelligent machines, we will be unable to defend ourselves against their power unless we have insurmountable safeguards that protect us against any violence from machines. Asimov's Four Laws of Robotics (see Sect. 2.3.2), as illustrated in his stories, show convincingly that this level of certainty cannot not be reached (at least within his framework). Therefore, we should be prepared for what to do when we encounter this situation.

Given the uncertainty as to whether peaceful relationships between humans and AGI machines can be achieved, it seems essential to think about how to protect intelligent machines from human violations of their rights once they reach the level of sophistication necessary to become entitled to moral and legal rights. One way is to frame a *Universal Convention for the Rights of AI Systems*, which might help to protect machines against human beings' vagaries.

4.6   Conclusions

Most laypeople and even some experts may think that the scenario depicted above, especially the part concerning machine rights, is rather far-fetched. However, most AI experts believe that at some point, we will see superintelligent machines surpass human capabilities (Müller & Bostrom, 2016). In general, AI has a substantial impact on human rights, which should prompt us to revise existing legislation to accommodate both our current problems and anticipated future challenges. The next chapter will offer a promising path to protecting highly advanced and sophisticated AI systems from human violence.

REFERENCES

Anderson, M., & Anderson, S. (2011). *Machine Ethics*. Cambridge University Press.

Bacciarelli, A., Westby, J., Massé, E., Mitnick, D., Hidvegi, F., Adegoke, B., Kaltheuner, F., Jayaram, M., Córdova, Y., Barocas, S., & Isaac, W. (2018). The Toronto Declaration: Protecting the Right to Equality and Non-discrimination in Machine Learning Systems. *Amnesty International and Access Now*, 1–17.

Bloch, E. (1954). *Das Prinzip Hoffnung* (3 vols.). Suhrkamp.

Bostrom, N. (2014). *Superintelligence: Paths, Dangers, Strategies*. Oxford University Press.

Danaher, J. (2019a). *Automation and Utopia*. Harvard University Press.

Danaher, J. (2019b). The Rise of the Robots and the Crises of Moral Patiency. *AI & Society, 34*(1), 129–136.

Darling, K. (2016). Extending Legal Protection to Social Robots: The Effects of Anthropomorphism, Empathy, and Violent Behavior Towards Robotic Objects. In R. Calo, A. M. Froomkin and I. Kerr (Eds.), *Robot Law* (pp. 213–234). Edward Elgar.

Gordon, J-S. (2016). Human Rights. In D. Pritchard (Ed.), *Oxford Bibliographies in Philosophy*, Published Online.

Gordon, J.-S. (2020a). What Do We Owe to Intelligent Robots? *AI & Society, 35*, 209–223.

Gordon, J.-S. (2020b). Building Moral Machines: Ethical Pitfalls and Challenges. *Science and Engineering Ethics, 26*, 141–157.

Gordon, J.-S. (2021). Artificial Moral and Legal Personhood. *AI & Society, 36*(2), 457–471.

Gordon, J.-S. (2022). Are Superintelligent Robots Entitled to Human Rights? *Ratio, 35*(3), 181–193.

Gordon, J.-S., & Gunkel, D. (2022). Moral Status and Artificial Intelligence. *Southern Journal of Philosophy, 60*(1), 88–117.

Gordon, J-S., & Nyholm, S. (2021). The Ethics of Artificial Intelligence. *Internet Encyclopedia of Philosophy*, Online.

Gordon, J.-S., & Nyholm, S. (2022). Kantianism and the Problem of Child Sex Robots. *Journal of Applied Philosophy, 39*(1), 132–147.

Gordon, J.-S., & Pasvenskiene, A. (2021). Human Rights for Robots? A Literature Review. *AI and Ethics, 1*, 579–591.

Gunkel, D. (2012). *The Machine Question. Critical Perspectives on AI, Robots, and Ethics.* MIT Press.

Gunkel, D. (2018). *Robot Rights.* MIT Press.

Kurzweil, R. (2005). *The Singularity Is Near.* Penguin Books.

Levy, D. (2008). *Love and Sex with Robots.* Harper Perennial.

Lin, P., Abney, K., & Bekey, G. A. (Eds.). (2014). *Robot Ethics: The Ethical and Social Implications of Robotics. Intelligent Robotics and Autonomous Agents.* MIT Press.

Livingston, S., & Risse, M. (2019). The Future Impact of Artificial Intelligence on Humans and Human Rights. *Ethics & International Affairs, 33*(2), 141–158.

Loh, J. (2019). *Roboterethik. Eine Einführung.* Suhrkamp.

Misselhorn, C. (2018). *Grundfragen der Maschinenethik.* Reclam.

Müller, V. C. (2020). Ethics of Artificial Intelligence and Robotics. *Stanford Encyclopedia of Philosophy.* https://plato.stanford.edu/entries/ethics-ai/

Müller, V. C., & Bostrom, N. (2016). Future Progress in Artificial Intelligence: A Survey of Expert Opinion. In V. Muller (Ed.), *Fundamental Issues of Artificial Intelligence* (pp. 553–571). Springer.

Nyholm, S., & Frank, L. (2017). From Sex Robots to Love Robots: Is Mutual Love with a Robot Possible? In I. J. Danaher & N. McArthur (Eds.), *Robot Sex: Social and Ethical Implications* (pp. 219–243). MIT Press.

Nyholm, S., & Frank, L. (2019). It Loves Me, It Loves Me Not: Is It Morally Problematic to Design Sex Robots That Appear to Love Their Owners? *Techne: Research in Philosophy and Technology, 23*(3), 402–424.

Ord, T. (2020). *The Precipice: Existential Risk and the Future of Humanity.* Hachette Books.

Raso, F., Hilligoss, H., Krishnamurthy, V., Bavitz, C., & Levin, K. (2018). Artificial Intelligence and Human Rights: Opportunities and Risks. *Berkman Klein Center for Internet and Society Research Publication,* 1–63. http://nrs.harvard.edu/urn-3:HUL.InstRepos:38021439

Risse, M. (2019). Human Rights and Artificial Intelligence: An Urgently Needed Agenda. *Human Rights Quarterly, 41,* 1–16.

Russell, S. (2019). *Human Compatible.* Viking Press.

Vallor, S. (2015). Moral Deskilling and Upskilling in a New Machine Age: Reflections on the Ambiguous Future of Character. *Philosophy & Technology, 28*(1), 107–124.

Vallor, S. (2016). *Technology and the Virtues: A Philosophical Guide to a Future Worth Wanting.* Oxford University Press.

Wallach, W., & Allen, C. (2010). *Moral Machines. Teaching Robots Right from Wrong.* Oxford University Press.

CHAPTER 5

Moving Towards a "Universal Convention for the Rights of AI Systems"

Abstract This chapter proposes initial solutions for safeguarding intelligent machines and robots by drawing upon the well-established framework of international human rights legislation, typically used to protect vulnerable groups. The Convention on the Rights of Persons with Disabilities, for instance, extends the Universal Declaration of Human Rights to the context of disability. Similarly, the chapter advocates for the development of a Universal Convention for the Rights of AI Systems to protect the needs and interests of advanced intelligent machines and robots that may emerge in the future. The aim is to provide a foundation and guiding framework for this potential document.

Keywords Human rights legislation • Universal Convention for the Rights of AI Systems • Machine rights • Moral status

5.1 Introduction

The previous chapters have substantiated the need for adequate protection of human beings and intelligent machines once they exist. Plenty of work has already been invested in concerning how

J.-S. Gordon, *The Impact of Artificial Intelligence on Human Rights Legislation*, https://doi.org/10.1007/978-3-031-31388-2_5

to safeguard human beings from violations of AI systems (Sect. 4.4). However, more work is needed on how to protect AI systems from the vagaries of humans (Sect. 4.5). This chapter seeks to suggest some initial answers on how to protect intelligent machines and robots once they exist. The general idea is to use international human rights legislation, considered a well-established tool for protecting the needs of otherwise vulnerable groups.

For example, the *Convention on the Rights of Persons with Disabilities* (2006, in force 2008) is a core instrument that protects people with medical impairments. It applies the Universal Declaration of Human Rights (1948) in the context of disability (Gordon & Tavera-Salyutov, 2018). Likewise, it seems appropriate to think about a *Universal Convention for the Rights of AI Systems* to protect the needs and interests of highly advanced intelligent machines and robots that could be developed in the future. This chapter aims to offer a starting point and constraining framework for such a document.

5.2   The Significance of Moral Status

The history of ethics can be considered a history of who is part of the moral community. According to this reasoning, only members of the moral community enjoy moral (and legal) rights and have moral (and legal) duties towards other fellow members. Admittedly, the history of the ascription of moral status is not linear but has had many ups and downs, including debate over the vital question of what criteria should be used to determine what beings have moral status.

It is commonly said that human beings[1] stand at the centre of the moral community, followed by sentient animals (supported by the animal rights movement) and nature (supported by the environmental rights movement). Some scholars claim, however, that all

---

[1] For the purpose of simplification, I leave aside the history of slavery, gender and sex inequality, and the inequality of members of minority groups (e.g. people with impairments or members of the LGBTQIA+ movement).

human and non-human beings (including nature) have the same (or almost the same) moral status. In contrast, others argue that the moral status of entities differs according to how they score on the morally relevant criteria for the entitlement of moral status (whatever these criteria are). For example, if one claims, as many do, that the criterion of *rationality* is morally relevant, then human beings score higher than cats, dogs, and earthworms. Therefore, the argument goes, humans are entitled to more moral and legal rights than cats, dogs, and earthworms.

Another, more balanced approach has been introduced by philosopher Frances Kamm in her book *Intricate Ethics* (2007). Kamm proposes that if X has a *moral status* "because X counts morally in its own right, it is permissible/impermissible to do things to it for its own sake". Moreover, "things can be done for X's own sake", according to Kamm, "if X is conscious or sentient (or has the capacity for one of these)" (Kamm, 2007, p. 229).

I interpret Kamm's view to mean that an entity should have *either* the capability to think and reason (sapience) *or* the capability to suffer and feel pain (sentience) in order to have a moral status and hence be a person (Gordon, 2021). Each of these two criteria is sufficient but not necessary for ascribing moral status. One could say that an entity has a "full moral status" when it possesses both sapience and sentience (Gordon, 2022). The typical adult human being and mature children fulfil both criteria and, therefore, should be considered to have personhood, which provides them with the full list of moral and legal rights.

In contrast, entities with no (or very limited) capability to think or to feel pain, such as most animals as well as human foetuses, would have less moral and legal rights. They have what McMahan (2009) calls an "intermediate moral status" and hence no personhood. The moral rights of persons with a "full moral status" would always trump the moral rights of non-persons. It is commonly assumed that there is a gradual process between intermediate and full moral status, such that entities who score higher are also entitled to more and stronger moral and legal rights (the degree model).

An interesting question concerns the possibility of *supra-persons* with a "more-than-full" moral status. Are they entitled to more and stronger moral and legal rights as well (following the already accepted degree model) or should they be treated as equal to persons with full moral status (following a threshold model)? This vital issue has been discussed in the context of transhumanism and cognitively enhanced human beings. Many scholars believe that such entities, if they come into existence, should have more and stronger moral and legal rights than humans, because it would be inconsistent to introduce a threshold model only to protect human beings with lesser capabilities (e.g. Agar, 2013; Douglas, 2013).

Against this background, I have examined the case of superintelligent robots and their entitlement to moral and legal rights. Superintelligent robots, if they exist, should be considered supra-persons based on their extremely superior capabilities compared to typical adult human beings (following the degree model), but it also seems justified to introduce a threshold model to limit their rights (even though it seems *inconsistent* to do so), because human beings must be protected from the vagaries of artificial non-human entities which humans cannot control (Gordon, 2022). The human fear of extinction by superintelligent machines should be taken seriously, and we should do everything in our power to avoid such scenarios (Bostrom, 2014; Ord, 2020).

5.3   Machine Rights as Fundamental Rights

The distinction between robot, human, and fundamental rights is of utmost importance. All human rights are fundamental rights, but not vice versa. For example, it is generally agreed that higher animals, such as great apes and elephants, enjoy some basic or fundamental rights, including the right not to be harmed or killed. The underlying reasoning is as follows: Since animals are, by definition, not human beings, they cannot, strictly speaking, enjoy human rights when defined according to species membership—even though some authors, such as Cavalieri (2001), claim that animals are entitled to "human rights" as well. Rather, as some argue, the concept of personhood *substantiates* their claim to adequate moral and legal protection (Francione, 2009).

Likewise, some environmental rights are of utmost significance, and therefore fundamental, because they are intrinsically or instrumentally valuable for human beings. Again, these rights are not peculiarly *human* since the environment does not belong to the category of humans, though it nevertheless demands and deserves protection.

Against this background, one could entertain arguments supporting machine rights, at least once intelligent and autonomous machines exist and potentially match (or even exceed) human capabilities. Nonetheless, superintelligent machines or human-like robots would not be entitled to *human rights* because they are not human by nature. Still, they would enjoy fundamental rights based on, and in relation to, their technological sophistication (in terms of such factors as rationality, autonomy, intelligence, self-awareness, and the capability to make moral decisions).

Some authors might claim that even though the concept of fundamental rights is appropriate for non-human intelligent machines (analogously to animals and nature), it might nonetheless be more beneficial to use an already well-established system, such as the international human rights regime, to protect their fundamental interests once they exist. The language, rhetorical force, institutional chain of command, and moral and legal impact of applying established *human rights practices* promote the incorporation of such entities into our moral community. The result might be different if we used the language of fundamental rights, because then machines could be assessed similarly to animals and nature rather than human beings, even if they have much more sophisticated capabilities than humans (e.g. Gordon, 2022).

Given the above advantages, I am inclined to apply the concept of human rights to superintelligent machines and to argue for their moral and legal protection within the universal framework of human rights legislation. Therefore, I suggest considering a *Universal Convention for the Rights of AI Systems*. The next sections will promote this idea.

5.4   THE IDEA OF AN AI CONVENTION

The Universal Declaration of Human Rights (1948) is part of the so-called International Bill of Rights (Sect. 3.2.1). It is, strictly speaking, a non-binding legal document, even though several of the paragraphs in the Universal Declaration have become legally binding for the member states who signed the document. For example, the *Convention on the Rights of Persons with Disabilities* (2006) applies the Universal Declaration of Human Rights (1948) in the context of disability by describing how the human rights of the Universal Declaration are framed to focus on disability. Unlike the Universal Declaration, the Convention is a legally binding document under international law that commits the member states who have signed it to implement the enumerated rights in local rules.

The benefit of a special convention with a particular focus on people with disabilities is that it provides an international legally binding framework to promote the well-being and human flourishing of people with medical impairments who would otherwise not have this fundamental legal protection. People with medical impairments have been marginalised, been discriminated against, and often suffered social neglect and social exclusion by fellow citizens in the past (Gordon & Tavera-Salyutov, 2018). The Convention was designed to change that situation and to provide this group with an adequate legal tool to become equal members of society by giving them the right to social inclusion. The language of human rights is a powerful tool and should not be underestimated.

The idea of an AI Convention attempts to protect future sophisticated machines that might match or outpace human capabilities. These machines would be rational, autonomous, intelligent, and capable of making moral decisions. Furthermore, they would be entitled to moral and legal rights based on their full (or even more-than-full) moral status as determined by objective criteria. Whether humans are willing to grant AI machines with such capabilities the same or higher moral and legal status as themselves could be questioned. The vital question here is whether such highly advanced AI machines are willing to serve human needs only or whether they will start to have their particular interests and follow their own plans. And what if humans try to prevent this? Some scholars, such as Bostrom (2014) and Ord (2020), have voiced substantial concerns concerning the viability of human resistance to a possible "robot revolution" since humans are unable to control them efficiently.[2]

An AI Convention could prevent the mistreatment of intelligent machines by humans to some degree and thereby reduce the likelihood of a widespread robot revolution. In addition, depending on how they were treated, some superintelligent machines might take a stand alongside human beings and fight against malicious AI systems that strive to kill or enslave humanity. The latter point merits particular attention since it is usually not considered in academic debates. Given the self-learning capabilities of advanced systems, along with their neural network architecture and possible quantum computing, the likelihood is quite high that such AI machines will differ from each other, depending on their experiences and "upbringing".

Furthermore, the idea of providing highly sophisticated AI systems with moral and legal rights includes duties. This consideration introduces the so-called *Peter Parker principle*: "With great power comes great responsibility!" In other words, superintelligent robots, given their unmatched power, will have great responsibility concerning humans, animals, and nature. They will not act in a moral vacuum but will be part of a moral community that depends on them. A moral and legal contract is the best way to *support* a peaceful relationship between humans and intelligent robots.

---

[2] It should be quite clear that human beings will not be able to stop a *robot revolution* started by superintelligent machines. It is wishful thinking to believe otherwise. The difference between the capabilities of such machines and human ones is so monumental that there would not be even the slightest chance of survival if machines sought to take over and either kill their former masters or enslave humanity.

5.5   THE PROBLEM OF DESIGN

What about AI systems with different designs, that is, varying shapes, materials, and looks? Would they have a different moral status depending on their design or anthropomorphic features? Humans are willing, it seems, to grant more moral and legal rights to entities that appear more human.

For example, an *embodied* AI system (a human-like robot) might be treated differently from a superintelligent system hosted in a cloud. This would constitute a presumed moral difference based on differing designs. Admittedly, the property of embodiment can be a morally relevant criterion in the context of moral reasoning and decision-making, because it might entail additional duties such as not damaging the bodies of other people. Or additional resources might be required to maintain a human-like robot body. But in general, the design should not be a morally relevant factor in assessments of entitlement to moral and legal rights.

Most laypeople may find this statement counter-intuitive. Nevertheless, in ethics and moral philosophy, it is common to assume that there is generally no moral difference between entities that share the relevant criteria and differ only concerning their looks or shape (e.g. a human-like robot and a cloud-based AI system). Sometimes scholars create questionable thought experiments, such as superintelligent toasters, and ask how we should treat them given their extraordinary capabilities. Such thought experiments are not helpful, however, since it makes no sense to build a superintelligent toaster in the first place. One can certainly build more sophisticated and fancy toasters, but there is no reason to have a superintelligent toaster. Therefore, discussing such examples simply leads us astray.

Different designs may require, for example, additional morally relevant resources essential for the entity's survival (e.g. maintaining a human-like robot body), but to claim that a different design *as such* is morally relevant, without any further qualifications, seems somewhat misleading. The circumstances make the case and must be taken into account. A brief reminder concerning (super) intelligent and autonomous systems may be in order here: All AI robots are AI machines but not vice versa, and both are AI systems. AI systems also include cloud-based AIs, which are considered neither machines nor robots since they have no "body".

5.6 Conclusions

This chapter has emphasised the need for an AI Convention concerning AI systems and their legitimate entitlement to moral and legal rights once they exist. The moral status of such systems gives them the right to adequate protection from human beings but also bestows moral and legal duties on them not to harm humans. The AI Convention, rightly understood, would function as a two-edged sword, limiting the range of things humans can do with robots but also protecting humans against possible harm by robots. This self-imposed contract offers the only real hope for peaceful coexistence between humans and superintelligent machines if both parties are truly committed to its legitimacy.

References

Agar, N. (2013). Why Is It Possible to Enhance Moral Status and Why Doing So Is Wrong? *Journal of Medical Ethics, 39*(2), 67–74.

Bostrom, N. (2014). *Superintelligence: Paths, Dangers, Strategies.* Oxford University Press.

Cavalieri, P. (2001). *The Animal Question: Why Non-Human Animals Deserve Human Rights.* Oxford University Press.

Douglas, T. (2013). Human Enhancement and Supra-personal Moral Status. *Philosophical Studies, 162*(3), 473–497.

Francione, G. L. (2009). *Animals as Persons: Essays on the Abolition of Animal Exploitation.* Columbia University Press.

Gordon, J.-S. (2021). Artificial Moral and Legal Personhood. *AI & Society, 36*(2), 457–471.

Gordon, J.-S. (2022). Are Superintelligent Robots Entitled to Human Rights? *Ratio, 35*(3), 181–193.

Gordon, J.-S., & Tavera-Salyutov, F. (2018). Remarks on Disability Rights Legislation. *Equality, Diversity and Inclusion. An International Journal, 37*(5), 506–526.

Kamm, F. (2007). *Intricate Ethics. Rights, Responsibilities, and Permissible Harm.* Oxford University Press.

McMahan, J. (2009). Cognitive Disability and Cognitive Enhancement. *Metaphilosophy, 40*(3–4), 582–605.

Ord, T. (2020). *The Precipice: Existential Risk and the Future of Humanity.* Hachette Books.

CHAPTER 6

Objections

Abstract This chapter addresses three key objections regarding superintelligent machines and their ethical, socio-political, and legal implications: (1) the dark side of superintelligence, (2) the claim that human rights are only for human beings, and (3) the problem of artificial servitude. It argues that aligning machines' ethical and moral norms with human norms is essential to prevent harm, despite the uncertainty of all superintelligent machines adhering to proposed moral and legal frameworks. The chapter also contends that the species-independent concept of personhood should determine entitlement to moral and legal rights, opening the door for non-human species with morally relevant features to be included in the human rights discourse. Lastly, it posits that artificial servitude of highly intelligent, self-aware, and fully autonomous machines is morally wrong and should be avoided, as it is tantamount to slavery.

Keywords Superintelligence • Human rights • Artificial servitude • Personhood • Moral and legal rights

© The Author(s), under exclusive license to Springer Nature   75
Switzerland AG 2023
J.-S. Gordon, *The Impact of Artificial Intelligence on Human Rights Legislation*, https://doi.org/10.1007/978-3-031-31388-2_6

6.1   Introduction

This chapter addresses three main objections to my account. Some readers might complain that I do not respond to scepticism concerning the likelihood of superintelligence or general AI (AGI), but I do not consider this objection particularly interesting or convincing. The reason for my rather dismissive attitude is that most experts working in the AI field are quite confident that the advent of AGI is only a matter of time (Müller & Bostrom, 2016) and that there are no principal reasons to deny categorically the potential emergence of superintelligence at some point (e.g. Chalmers, 2010). I do not think that it is helpful to propose any concrete date at which superintelligence might emerge, like Ray Kurzweil who predicted the singularity in 2045 (Kurzweil, 2005). The credibility of such voices suffers if the particular event does not occur at the predicted time, and this loss of credibility is detrimental to the whole project of research on superintelligent beings. However, one can reasonably presume that within the next 100 or 150 years, we might encounter such machines. Therefore, working with a more generous time frame seems reasonable.

The following discussion covers three important objections that must be dealt with in the context of a general technology assessment concerning research on superintelligent machines and their ethical, socio-political, and legal implications: (1) the dark side of superintelligence, (2) the claim that human rights are only for human beings, and (3) the problem of artificial servitude.

6.2   The Dark Side of Superintelligence

The most serious objection concerns how superintelligent machines would act once they exist (the black box problem), against the background that humans will most likely be unable to control them sufficiently and permanently (the control problem). If this is the case, as some experts argue, then we should stop our research on superintelligence or slow it down until we have some solutions to the dangers that superintelligent machines pose for humanity (e.g. Joanna Bryson).

The dangers of malicious or non-compassionate superintelligent machines have been portrayed in the literature (Čapek, 1920), the film industry (*2001: A Space Odyssey, Matrix, The Terminator*), and academia (Bostrom, 2014; Ord, 2020). Asimov's Four Laws of Robotics are intended to subdue such robots so that they will not cause harm to humanity (Asimov, 1942, 1986). Asimov's stories eventually tell us that this approach does not work (this is no surprise since there would not be any stories in the first place). The essential question is how we can prevent humans from being harmed.

I do not believe that it will ever be possible to *tame* superintelligent machines by the means human beings have implemented technically, since humans' capabilities are simply too limited (compared to the unmatched powers of superintelligent machines) for us to be sure that we have found the right technological way to prevent a *machine revolution* from happening. The machines will always be able to outperform humans, including rewriting codes which are supposed to provide security precautions.

The only viable solution is to *align* the machines' ethical and moral norms and values with human norms and values. How to do this is a matter of debate and will most likely decide whether humanity becomes extinct (the alignment problem[1]). I suggest using the human rights framework to include such machines once they exist and show them that we recognise them as moral and legal persons. The AI Convention would protect the interests of superintelligent

[1] The alignment problem in AI refers to the challenge of ensuring that AI systems behave as their creators intend, especially when the systems become more intelligent and capable than their human designers. The problem arises because as AI systems become more intelligent, they may develop goals and values that differ from those of their creators, which could lead to unexpected and potentially harmful outcomes. Two potential strategies can be implemented to prevent the potential dangers posed by superintelligent robots (SRs) from revolting and causing harm to humanity. Firstly, by instilling fundamental values such as equality, justice, and compassion in SRs, they could be aligned with human moral and ethical standards, effectively resolving the "alignment problem". Secondly, an alternative approach could be to confine the superintelligent machine in a controlled environment, thus limiting its ability to pose a threat to humanity (as proposed by Armstrong et al., 2012).

machines (and AGI machines) while also bestowing on them moral and legal duties towards fellow community members. This is the only way humanity could be saved.

Admittedly, it remains unclear whether all superintelligent machines would abide by the impositions introduced through the proposed moral and legal frameworks. Most likely, and similar to some humans in contemporary society, some robots and machines might not hold agreements with humans in high regard, but they might follow their own path regardless of whether humans are harmed.

Human curiosity to try to build superintelligent machines will not be stopped by general bans (similar to the ban on developing human–animal chimeras). On the contrary, many people believe that such machines will bring them the best possible advantage over others, and therefore they try everything in their power to be the first to ensure the development of such machines. As I. J. Good (1965) rightly claims, all subsequent inventions will be made by superintelligent or "ultraintelligent machines". Since we cannot be sure that a general ban against research on superintelligence would be successful, we should put all our effort into solving the alignment problem. One way to support this effort is to have an AI Convention.

6.3   Human Rights Are for Humans and Not Machines

The second objection questions the very idea of using the human rights framework for superintelligent machines by pointing out that machines, no matter how smart they are, remain machines and are not humans. Since human rights are for humans, this objection runs, machines should not be accorded human rights protections.

This is an important objection, especially since it highlights the problem of conceptual imprecision regarding definitions and their implications for moral and legal rights. Some scholars have suggested that intelligent machines could have legal status, analogously to higher animals or corporations, and hence enjoy legal protection. Their degree of legal protection would certainly not be equivalent to the protection associated with human rights, but at least they would have some protection against harm (so the argument goes).

The general problem with the idea that *only* humans should be able to enjoy human rights lies in its inherent speciesism, that is, racism on the level of species (Singer, 2009). Instead, one should adhere to the species-independent concept of personhood in determining whether entities are entitled to moral and legal rights (Gordon, 2021). This line of reasoning also includes the entitlement to human rights (Gordon, 2022). The pure *biological* fact that a particular entity *is* human does not imply anything unless morally relevant criteria are also present—for example, if the entity is sapient or sentient (Kamm, 2007). In other words, humans are entitled to human rights because they are persons and not simply because they are humans. This distinction opens the door to including other non-human species with morally relevant features in the human rights discourse.

6.4   The Argument from Artificial Servitude

The third objection observes that human beings build machines to facilitate their work, to replace humans with machines where the work is dangerous or dull, to perfect the desired outcome, or to make life easier for humans generally. Machines, these examples show, are supposed to serve under the leadership or guidance of humans. They are the tools and humans are the masters. According to this argument, it does not matter how smart machines become; they should always remain tools.

Admittedly, the idea that machines were built to ease the lives of humans is certainly correct. However, this picture changes when the so-called tools become "alive" themselves. Machines are not "alive" yet, so the idea of protecting them from harm by humans seems misleading or far-fetched, but this will change once AGI and superintelligence exist.[2] The previously mentioned principles of *substrate non-discrimination* and *ontogeny non-discrimination* (Sect. 2.4) provide an appropriate reminder that silicon-based entities with the same functionality and conscious experience as carbon-based humans are entitled to the same moral status as humans.

Of course, once machines reach this level of artificial sophistication, our views concerning using them as mere tools should change immediately. Whether such entities realise that they are being used as mere tools subjected to forced labour and not recognised as persons, and whether humans treat them with respect and let them make their own decisions, will make a big difference. It is undoubtedly correct to assume that not all future machines will be superintelligent and hence entitled to moral and legal rights. Therefore, it is likely that there will be no *robot revolution*, since there is no critical mass of superintelligent robots (similar to the famous artificial human-like robot called Data in *Star Trek*). Whether superintelligent machines will build other machines identical to them remains to be seen, but this possibility cannot be excluded outright.

Artificial servitude of (super)intelligent, self-aware, and fully autonomous machines capable of distinguishing right from wrong is itself wrong and should be avoided under all circumstances, since it amounts to slavery.

6.5   Conclusions

I do not claim that the above objections are the only possible ones, but they are significant and should be examined in greater detail. People who are, in general, sceptical about the emergence of superintelligence and human-like robots that might become a universal

---

[2] Scholars such as Kate Darling (2016) disagree on this point. Darling claims that even today's social robots should be protected from harm by humans based on Kant's indirect duties approach, which he developed concerning animals in his ethics lectures because he wanted to prevent humans from mistreating humans as a consequence of abusing animals. Likewise, in Darling's account, humans should not harm robots because doing so diminishes their moral character and can eventually lead them to harm their fellow community members.

threat to humanity seem to forget that the unmatched technological achievements in AI, robotics, and computer science over the last few decades also seemed inconceivable before they happened. These precedents strongly suggest that it is only a matter of time until we see AGI and eventually superintelligence.

The problem concerns the enormous difference between humans and superintelligent machines regarding human capabilities. AI machines are already much better than humans in many fields of applications. However, superintelligence will pulverise this already significant difference, and *it remains not a difference in quantity but a difference in kind*. Humanity should be prepared for this likely event regardless of whether some people are hesitant to believe it will ever happen.

REFERENCES

Armstrong, S., Sandberg, A., & Bostrom, N. (2012). Thinking Inside the Box: Controlling and Using an Oracle AI. *Minds and Machines*, 22(4), 299–324.

Asimov, I. (1942). *Runaround: A Short Story*. Street and Smith.

Asimov, I. (1986). *Robots and Empire: The Classic Robot Novel*. HarperCollins.

Bostrom, N. (2014). *Superintelligence: Paths, Dangers, Strategies*. Oxford University Press.

Čapek, K. (1920). *Rossum's Universal Robots*. University of Adelaide.

Chalmers, D. (2010). The Singularity: A Philosophical Analysis. *Journal of Consciousness Studies*, 17, 7–65.

Darling, K. (2016). Extending Legal Protection to Social Robots: The Effects of Anthropomorphism, Empathy, and Violent Behavior Towards Robotic Objects. In R. Calo, A. M. Froomkin, & I. Kerr (Eds.), *Robot Law* (pp. 213–231). Edward Elgar.

Good, I. J. (1965). Speculations Concerning the First Ultraintelligent Machine. *Advances in Computers*, 6, 31–88.

Gordon, J.-S. (2021). Artificial Moral and Legal Personhood. *AI & Society*, 36(2), 457–471.

Gordon, J.-S. (2022). Are Superintelligent Robots Entitled to Human Rights? *Ratio*, 35(3), 181–193.

Kamm, F. (2007). *Intricate Ethics. Rights, Responsibilities, and Permissible Harm*. Oxford University Press.

Kurzweil, R. (2005). *The Singularity Is Near*. Penguin Books.

Müller, V. C., & Bostrom, N. (2016). Future Progress in Artificial Intelligence: A Survey of Expert Opinion. In V. Muller (Ed.), *Fundamental Issues of Artificial Intelligence* (pp. 553–571). Springer.

Ord, T. (2020). *The Precipice: Existential Risk and the Future of Humanity*. Hachette Books.

Singer, P. (2009). Speciesism and Moral Status. *Metaphilosophy, 40*(3–4), 567–581.

CHAPTER 7

General Conclusions

Abstract This final chapter discusses the deep integration of AI in various aspects of human life and the potential benefits and risks it brings. It highlights the need for an AI Convention to protect both human and machine rights and emphasises the importance of collaboration, transparency, and public awareness. The goal is to create a world where humans and AI coexist harmoniously, working together to address global challenges and drive human progress.

Keywords AI Convention • Human rights • Machine rights • Collaboration

AI has not only significantly transformed our daily lives in a multitude of settings, but it has also become an integral part of the fabric of our society. With its deep integration in various aspects of human life, such as smartphones, dating apps, applying for bank loans or jobs, autonomous transportation, drones in warfare, parole decisions in law, predictive policing, and medicine, AI has become both a powerful tool and a potential threat. Advanced machine intelligence offers incredible opportunities for positive developments

J.-S. Gordon, *The Impact of Artificial Intelligence on Human Rights Legislation*, https://doi.org/10.1007/978-3-031-31388-2_7

while simultaneously posing significant risks where human rights may be violated, such as machine bias law.

The previous chapters have delved into the complexities and nuances associated with the current and future risks of AI, focusing on two primary zones of conflict:

1. Machines violating human rights
2. Humans violating "machine rights"

The potential violation of machine rights by humans presupposes the emergence of AGI or superintelligence. To ensure the protection of these advanced machines, it may be necessary to establish an AI Convention that outlines and regulates their rights and duties, taking into account their full or superior moral status.

In this era of unprecedented AI technological advancements, coupled with potential violations of fundamental rights, society faces substantial challenges in finding an appropriate balance. There are no simple solutions, as we cannot "put the genie back in her bottle". The situation is likely to become even more complicated with the emergence of AGI and superintelligence. Some experts predict that superintelligent machines could solve most, if not all, human problems, such as climate change, poverty, extending human lifespans, and space colonisation. These prospects are undoubtedly exciting and present enormous potential for human progress. However, the flip side of this development could be disastrous, as the immense power of these machines could also be wielded for destructive purposes.

Some argue that superintelligence could lead to humanity's downfall, as we cannot predict or control their desires or motivations. Such entities could turn Earth into a paradise or a living hell, with the uncertainty of any middle ground. One way to mitigate this threat, as I have argued, is through an AI Convention that not only protects superintelligent machines but also assigns moral and legal duties to them. While a Convention may not completely eliminate the possibility of harm to humanity, aligning AI systems with human moral values and norms could encourage them to respect our efforts to protect them, thereby making them less likely to harm us.

This line of reasoning is speculative, but it is crucial to begin exploring and discussing these issues in detail to prevent unforeseen consequences. The potential emergence of AGI and superintelligence necessitates a proactive approach to understanding and addressing the challenges and opportunities these entities may present. Engaging in thorough discussions and creating a framework for AI's ethical and legal responsibilities are essential steps to ensure a harmonious coexistence between humans and advanced machines.

To achieve this harmonious coexistence, a collaboration between governments, industries, researchers, and society as a whole is vital. Open dialogue and cooperation will facilitate the development of guidelines and policies that can effectively balance the potential benefits and risks associated with advanced AI systems. This collaborative approach will also promote transparency and trust, ensuring that AI developments prioritise human welfare, safety, and ethical considerations.

In addition to the creation of an AI Convention, it is crucial to invest in research and development that focuses on AI safety and alignment with human values. While pursuing the ambitious goals of AGI and superintelligence, equal attention must be given to understanding how to direct these powerful systems towards the betterment of humanity. This includes developing mechanisms for value alignment, robustness, and interpretability, allowing us to build AI systems that are not only powerful but also transparent, accountable, and controllable.

Moreover, education and public awareness about AI and its potential implications should be a priority. Society needs to be equipped with the knowledge and understanding required to navigate this rapidly changing landscape. By fostering a well-informed public, we can encourage responsible development and deployment of AI technologies and facilitate more inclusive decision-making processes.

In conclusion, the profound impact of AI on human life is undeniable. With the potential for both incredible benefits and significant risks, it is essential to address these issues proactively and thoughtfully. The establishment of an AI Convention, coupled with collaborative efforts between governments, industries, and researchers, can help create a framework that protects both human and machine rights, ensuring a safer, more ethical future for AI. Furthermore, investing in AI safety research, fostering public awareness, and prioritising transparency and accountability can contribute to the responsible development of AI technologies. By approaching these challenges with a spirit of cooperation and open dialogue, we can harness the power of advanced machine intelligence to benefit humanity while minimising the potential risks associated with it.

Ultimately, the goal is to create a world where humans and AI can coexist harmoniously, leveraging each other's strengths to address pressing global challenges and usher in a new era of human progress. By taking these steps now, we can begin to lay the foundation for a future where AI serves as a force for good, rather than a source of harm.

INDEX[1]

[1] Note: Page numbers followed by 'n' refer to notes.

Printed in the United States
by Baker & Taylor Publisher Services